STORING SPENT FUEL UNTIL TRANSPORT TO REPROCESSING OR DISPOSAL

The following States are Members of the International Atomic Energy Agency:

AFGHANISTAN
ALBANIA
ALGERIA
ANGOLA
ANTIGUA AND BARBUDA
ARGENTINA
ARMENIA
AUSTRALIA
AUSTRIA
AZERBAIJAN
BAHAMAS
BAHRAIN
BANGLADESH
BARBADOS
BELARUS
BELGIUM
BELIZE
BENIN
BOLIVIA, PLURINATIONAL
 STATE OF
BOSNIA AND HERZEGOVINA
BOTSWANA
BRAZIL
BRUNEI DARUSSALAM
BULGARIA
BURKINA FASO
BURUNDI
CAMBODIA
CAMEROON
CANADA
CENTRAL AFRICAN
 REPUBLIC
CHAD
CHILE
CHINA
COLOMBIA
CONGO
COSTA RICA
CÔTE D'IVOIRE
CROATIA
CUBA
CYPRUS
CZECH REPUBLIC
DEMOCRATIC REPUBLIC
 OF THE CONGO
DENMARK
DJIBOUTI
DOMINICA
DOMINICAN REPUBLIC
ECUADOR
EGYPT
EL SALVADOR
ERITREA
ESTONIA
ESWATINI
ETHIOPIA
FIJI
FINLAND
FRANCE
GABON
GEORGIA

GERMANY
GHANA
GREECE
GRENADA
GUATEMALA
GUYANA
HAITI
HOLY SEE
HONDURAS
HUNGARY
ICELAND
INDIA
INDONESIA
IRAN, ISLAMIC REPUBLIC OF
IRAQ
IRELAND
ISRAEL
ITALY
JAMAICA
JAPAN
JORDAN
KAZAKHSTAN
KENYA
KOREA, REPUBLIC OF
KUWAIT
KYRGYZSTAN
LAO PEOPLE'S DEMOCRATIC
 REPUBLIC
LATVIA
LEBANON
LESOTHO
LIBERIA
LIBYA
LIECHTENSTEIN
LITHUANIA
LUXEMBOURG
MADAGASCAR
MALAWI
MALAYSIA
MALI
MALTA
MARSHALL ISLANDS
MAURITANIA
MAURITIUS
MEXICO
MONACO
MONGOLIA
MONTENEGRO
MOROCCO
MOZAMBIQUE
MYANMAR
NAMIBIA
NEPAL
NETHERLANDS
NEW ZEALAND
NICARAGUA
NIGER
NIGERIA
NORWAY
OMAN
PAKISTAN

PALAU
PANAMA
PAPUA NEW GUINEA
PARAGUAY
PERU
PHILIPPINES
POLAND
PORTUGAL
QATAR
REPUBLIC OF MOLDOVA
ROMANIA
RUSSIAN FEDERATION
RWANDA
SAINT VINCENT AND
 THE GRENADINES
SAN MARINO
SAUDI ARABIA
SENEGAL
SERBIA
SEYCHELLES
SIERRA LEONE
SINGAPORE
SLOVAKIA
SLOVENIA
SOUTH AFRICA
SPAIN
SRI LANKA
SUDAN
SWEDEN
SWITZERLAND
SYRIAN ARAB REPUBLIC
TAJIKISTAN
THAILAND
THE FORMER YUGOSLAV
 REPUBLIC OF MACEDONIA
TOGO
TRINIDAD AND TOBAGO
TUNISIA
TURKEY
TURKMENISTAN
UGANDA
UKRAINE
UNITED ARAB EMIRATES
UNITED KINGDOM OF
 GREAT BRITAIN AND
 NORTHERN IRELAND
UNITED REPUBLIC
 OF TANZANIA
UNITED STATES OF AMERICA
URUGUAY
UZBEKISTAN
VANUATU
VENEZUELA, BOLIVARIAN
 REPUBLIC OF
VIET NAM
YEMEN
ZAMBIA
ZIMBABWE

The Agency's Statute was approved on 23 October 1956 by the Conference on the Statute of the IAEA held at United Nations Headquarters, New York; it entered into force on 29 July 1957. The Headquarters of the Agency are situated in Vienna. Its principal objective is "to accelerate and enlarge the contribution of atomic energy to peace, health and prosperity throughout the world".

STORING SPENT FUEL UNTIL TRANSPORT TO REPROCESSING OR DISPOSAL

INTERNATIONAL ATOMIC ENERGY AGENCY
VIENNA, 2019

COPYRIGHT NOTICE

All IAEA scientific and technical publications are protected by the terms of the Universal Copyright Convention as adopted in 1952 (Berne) and as revised in 1972 (Paris). The copyright has since been extended by the World Intellectual Property Organization (Geneva) to include electronic and virtual intellectual property. Permission to use whole or parts of texts contained in IAEA publications in printed or electronic form must be obtained and is usually subject to royalty agreements. Proposals for non-commercial reproductions and translations are welcomed and considered on a case-by-case basis. Enquiries should be addressed to the IAEA Publishing Section at:

Marketing and Sales Unit, Publishing Section
International Atomic Energy Agency
Vienna International Centre
PO Box 100
1400 Vienna, Austria
fax: +43 1 26007 22529
tel.: +43 1 2600 22417
email: sales.publications@iaea.org
www.iaea.org/books

© IAEA, 2019

Printed by the IAEA in Austria
March 2019
STI/PUB/1846

IAEA Library Cataloguing in Publication Data

Names: International Atomic Energy Agency.
Title: Storing spent fuel until transport to reprocessing or disposal / International Atomic Energy Agency.
Description: Vienna : International Atomic Energy Agency, 2019. | Series: IAEA nuclear energy series, ISSN 1995–7807 ; no. NF-T-3.3 | Includes bibliographical references.
Identifiers: IAEAL 19-01223 | ISBN 978–92–0–100719–3 (paperback : alk. paper)
Subjects: LCSH: Spent reactor fuels — Storage. | Nuclear fuels. | Nuclear industry — Safety measures.
Classification: UDC 621.039.54 | STI/PUB/1846

FOREWORD

One of the IAEA's statutory objectives is to "seek to accelerate and enlarge the contribution of atomic energy to peace, health and prosperity throughout the world." One way this objective is achieved is through the publication of a range of technical series. Two of these are the IAEA Nuclear Energy Series and the IAEA Safety Standards Series.

According to Article III.A.6 of the IAEA Statute, the safety standards establish "standards of safety for protection of health and minimization of danger to life and property". The safety standards include the Safety Fundamentals, Safety Requirements and Safety Guides. These standards are written primarily in a regulatory style, and are binding on the IAEA for its own programmes. The principal users are the regulatory bodies in Member States and other national authorities.

The IAEA Nuclear Energy Series comprises reports designed to encourage and assist R&D on, and application of, nuclear energy for peaceful uses. This includes practical examples to be used by owners and operators of utilities in Member States, implementing organizations, academia, and government officials, among others. This information is presented in guides, reports on technology status and advances, and best practices for peaceful uses of nuclear energy based on inputs from international experts. The IAEA Nuclear Energy Series complements the IAEA Safety Standards Series.

The lack of sufficient political will and public support presents significant challenges for implementing an end point for spent fuel management (SFM) — such as reprocessing or disposal — resulting in the need for longer storage periods. Given present and projected rates for the use of nuclear power, coupled with projections for reprocessing and disposal of spent fuel, it is expected that spent fuel storage periods may be extended for several decades.

It is imperative that States make the necessary political decisions to define and implement an end point for SFM needed for the responsible and sustainable use of nuclear power. This can eliminate unnecessary risks, including those associated with packaging spent fuel in the absence of acceptance criteria for reprocessing or disposal, as well as unnecessary costs associated with the maintenance of institutional controls and ageing management programmes for longer storage periods.

In the interim, it is essential to ensure safe, secure and effective storage of spent fuel under all foreseeable conditions. This publication explores approaches to achieve this objective given present unknown storage durations. It identifies issues and challenges relevant to the development and implementation of options, policies, strategies and programmes to accommodate the full range of future scenarios until sufficient reprocessing or disposal capacity becomes available. It is not intended to either facilitate or encourage the extension of spent fuel storage durations.

This publication is to help the nuclear industry to communicate the importance of a clear, credible and sustainable SFM strategy and will encourage decision makers to consider different approaches that may be useful in addressing the uncertainties resulting from an unknown storage duration and an undefined end point for SFM. This publication applies to States that operate or have operated power reactors and research, material testing or isotope production reactors, and to newcomers considering the introduction of nuclear power. It is recognized that specific solutions will vary depending on the need of the State and that the approaches presented here will be adapted accordingly.

The IAEA wishes to express its appreciation to everyone who took part in the preparation of this publication, and in particular to B. Carlsen (United States of America), who chaired the technical meetings and consultant meetings, and coordinated and contributed to the drafting and review of this publication. The IAEA officers responsible for this publication were A. Bevilacqua and A. González-Espartero of the Division of Nuclear Fuel Cycle and Waste Technology.

CONTENTS

SUMMARY

Spent fuel storage periods well beyond those originally foreseen are a reality. This publication offers several ideas and approaches that may be considered to address the increasingly longer storage times. The aim is to raise awareness, encourage dialogue and provide ideas on how to manage spent fuel. Key messages include the following:

— Delays in reprocessing or disposal could result in spent fuel being stored for 100 years or longer. Safe, secure and effective storage of spent fuel manages fuel degradation while preserving future fuel cycle options.
— Ageing management programmes apply engineering, operations and maintenance actions to ensure safety is maintained during storage, future handling and transport.
— Site selection and facility and equipment design can significantly reduce the risks and costs of spent fuel storage over longer periods.
— Spent fuel storage configurations can be selected to accommodate uncertain storage periods, to facilitate ageing management and to provide flexibility needed to accommodate the uncertainty of future end points, such as reprocessing or disposal.
— By considering multiple licence renewals, regulatory frameworks can be designed to ensure safe storage until an acceptable end point is achieved.
— Safety can be assured by maintaining shielding, containment, decay heat removal and criticality control. Navigating the complexity of societal beliefs and values, as well as political systems, has proven to be a greater challenge for the management of spent fuel than maintaining its safety and security or addressing the technical and economic aspects.
— Sustainable spent fuel management requires policies and strategies to provide a clear, consistent and stable direction because they drive the need for spent fuel storage as well as the available options and timing for achieving an acceptable end point. Unless States address spent fuel reprocessing and disposal on a sufficient scale to accommodate their spent fuel discharges, then storage for longer and longer periods becomes the de facto end point — which is not considered to be consistent with the responsibility to protect human health and the environment.

An effective, periodic licence renewal process can ensure effective ageing management and strong institutional control. Hence, spent fuel can be safely and securely stored for as long as it may be necessary until transport for reprocessing or disposal. However, the risks and costs of storing the growing inventory of spent fuel will continue to increase; and in the absence of an end point, it will eventually become a significant societal burden.

1. INTRODUCTION

1.1. BACKGROUND

The nuclear fuel cycle encompasses all operations associated with the production of nuclear energy, from uranium exploration and mining to extraction from the earth to the manufacture of nuclear fuel to be used in the production of electricity, and ends when the resulting spent fuel or high level waste (HLW) from reprocessing reaches a suitable endpoint (see Refs [1–4]). The presently accepted endpoint is when spent fuel and HLW is safely placed into the ground in a suitable geologic repository. Worldwide, about 10 000 t HM of spent fuel is discharged from nuclear power plants in 30 IAEA Member States each year (see Fig. 1), while a few Member States reprocess spent fuel with a combined annual capacity of 4800 t HM (of which not all is operational) [5]. Owing to the growing demand for clean, reliable energy, several States have committed to expanding their nuclear capacity, thus further increasing the spent fuel to be stored and emphasizing the need to define and implement the spent fuel end point[1] (i.e. reprocessing or disposal).

Some Member States designate spent fuel for disposal, while others reprocess it to recover fuel material. Policy changes in some States have resulted in such designations being changed from one to the other. Until national policy is decided and implemented, stored spent fuel cannot begin moving to reprocessing or disposal. Because sufficient worldwide reprocessing and disposal capacity are decades away, spent fuel and HLW will continue to accumulate at storage facilities, and licence periods for spent fuel storage facilities will need to be renewed, perhaps multiple times. Although HLW storage is not within the scope of this publication, many of the issues presented still apply.

The dashed lines in Fig. 1 indicate flow restrictions that result in increasing inventories in storage. States can estimate spent fuel storage capacity and duration by substituting appropriate mass flows and by considering the capacity, projected start date, and confidence level for spent fuel reprocessing and disposal operations. The uncertainty in the starting date for deep geological repository operations means that even States that have decided to phase out nuclear power still have to manage their existing spent fuel inventory for time periods that cannot be confidently defined.

A defined end point, such as reprocessing or disposal, depends upon future funding, legislation, licensing and other conditions that cannot be predicted with certainty. This publication explores how this uncertainty can be considered in design, licensing, management strategy and policy decisions.

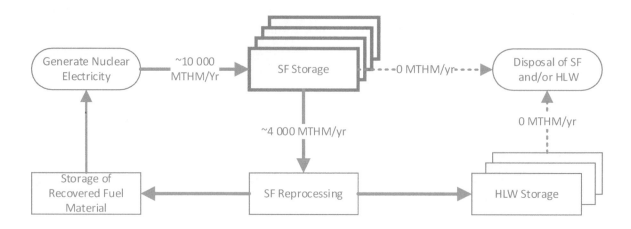

FIG. 1. Spent fuel storage.

---

[1] The end point is defined as the state of radioactive material in the final stage of its management, in which the material is passively safe and does not depend on institutional control.

1.2. OBJECTIVE

This publication provides information for key policy and decision makers responsible for establishing policies and programmes for spent fuel management (SFM) and identifies technical and non-technical considerations necessary when extending storage. It acknowledges the uncertainty in determining the length of spent fuel storage periods and helps to identify options and trade-offs relevant to the associated technical and regulatory uncertainties.

1.3. SCOPE

Several topics considered important for safely, securely and effectively extending spent fuel storage are addressed, and possible paths for avoiding or mitigating unnecessary risks and costs are presented. Although this publication primarily focuses on the storage of spent fuel from nuclear power reactors in commercial operation, the principles described are equally applicable to spent fuel from research reactors, material test reactors and isotope production reactors. The storage of HLW is not within the scope of this publication; however, many of the same issues apply. Guidance provided here, describing good practice, represents expert opinion but does not constitute recommendations made on the basis of a consensus of Member States.

1.4. STRUCTURE

Section 2 of this publication provides an overview of spent fuel storage safety and the sustainability of SFM. Section 3 outlines ageing management programmes and closing knowledge gaps. Section 4 presents considerations in the design and siting of future spent fuel storage systems, and Section 5 describes spent fuel storage configurations. Sections 6 and 7 examine regulatory and policy considerations, respectively, and Section 8 concludes with other key considerations.

2. OVERVIEW

Past assumptions of available capacity for spent fuel reprocessing and disposal have often been wrong, resulting in missed opportunities to implement policies and strategies. For example, dry cask storage systems (DCSSs) were originally conceived to free up space in reactor spent fuel pools and to provide spent fuel storage of up to 20 years until sufficient reprocessing or deep geological disposal capacity became available. Hundreds of DCSSs are now employed throughout the world and will be relied upon for well beyond their originally envisioned design life.

On account of the ongoing public and political debate on nuclear energy and on the siting and licensing of deep geological repositories, storage periods and projected spent fuel inventories cannot be known with certainty. Moving forward despite this uncertainty requires confidence that spent fuel can be safely stored until sufficient spent fuel reprocessing or deep geological disposal becomes available. As a result, many Member States are reconsidering the public, political, technical and regulatory issues associated with extending storage periods (see Refs [6–8]).

Bridges, dams, and other common structures and equipment can fail with significant costs to life and property. Yet, despite this uncertainty, the risks are still accepted. The public understands and values the benefits of such structures and has sufficient confidence that the relevant safety concerns can be recognized and addressed before catastrophic failure. Essential equipment is maintained and, at some point, retired and replaced before failure occurs. Similarly, the question is not whether spent fuel can be safely stored but rather what is needed to provide sufficient confidence that age related degradation will be recognized and addressed to prevent unacceptable safety consequences regardless of the storage duration. The safety basis and SFM strategies need not depend upon an unknowable future. Strategies can be identified and the appropriate infrastructure developed to ensure safe spent fuel storage that is independent of time. The risks and uncertainties associated with extending spent fuel storage can never be completely eliminated. They can, however, be managed to ensure that the likelihood of an unanticipated occurrence is sufficiently low that its consequences are sufficiently mitigated to render the risk acceptably low.

2.1. EXTENDING SPENT FUEL STORAGE ONE STEP AT A TIME

Historically, spent fuel storage has been referred to using terms such as short term storage, interim storage, long term storage, extended storage or very long term storage. Implicit in each is an assumption about the storage period. Extending spent fuel storage for an additional, fixed licensing period can be considered without knowing a priori how many successive periods will be necessary until the spent fuel is transported to reprocessing or disposal.

Potential hazards, available technologies and applicable requirements can change over time [9]. By developing a technical and regulatory approach that allows successive renewals for as long as necessary, the need to predict or define an end date for the storage period can be avoided [9]. Hence, the scope of this publication is not limited to any specific storage duration. Rather, the uncertainty in storage periods is addressed by extending spent fuel storage one step at a time, until the end point — reprocessing or disposal — becomes available.

In theory, the number of licence renewals need not be limited as long as compliance with requirements is demonstrated. In practice, however, the accumulated cost of maintaining compliance (i.e. maintenance and upgrades) may eventually favour upgrading to new storage systems [9].

2.2. SPENT FUEL STORAGE SAFETY

Spent fuel has been safely stored for over 50 years, and the necessary safety principles are well understood. Emerging challenges can be identified and corrected before they become safety issues. IAEA Safety Standards Series No. SSG-15, Storage of Spent Nuclear Fuel [10], provides guidance for ensuring the safety of spent fuel storage [10]. As stated in para. 1.3 of SSG-15 [10], "The safety of a spent fuel storage facility, and the spent fuel stored within it, is ensured by: appropriate containment of the radionuclides involved, criticality safety, heat removal, radiation shielding and retrievability" during all normal, off-normal and design basis accident conditions, as well as addressing beyond design basis accident conditions. Although future requirements cannot be predicted with certainty, the following fundamental spent fuel storage safety functions will not change over increased storage periods and therefore it is reasonable to assume that they will underpin any future spent fuel storage regulation [9].

2.2.1. Containment

Containment prevents the release of radioactive material into the environment and is provided by the spent fuel cladding and the storage system (e.g. welded or bolted canister or cask). As storage times lengthen, the potential for materials degradation increases. As a result, it may be necessary to shift containment functions to components that can be more readily inspected and maintained.

2.2.2. Subcriticality

Subcriticality precludes an unplanned criticality event. Ensuring subcriticality often relies on geometry control. However, maintaining the geometry of spent fuel during longer storage periods could become more challenging as a result of materials degradation adversely affecting the structural integrity of the spent fuel and storage system components (e.g. basket, neutron absorbers, canisters or casks). Criticality control functions can be shifted to other structures, systems and components (SSCs) that can more readily be assured, inspected and maintained. These include controlling fissile content, constraining geometry, inclusion of neutron absorbers and preclusion of sufficient moderators.

2.2.3. Decay heat removal

Decay heat removal precludes loss of geometry, which can reduce the safety margins for other safety functions. Temperature limits prevent a loss of integrity of the cladding and other SSCs important to safety. Cladding integrity is important for preserving spent fuel assembly (SFA) geometry, which can affect safety margins for subcriticality, shielding and containment. Effective removal of decay heat is important because many of the phenomena which adversely affect spent fuel integrity are thermally activated. Decay heat removal during extended storage may be less challenging because the spent fuel becomes cooler over time. Low temperatures, on the other hand, might be an issue during transport after long term storage due to potential embrittlement of the cladding.

2.2.4. Shielding

Shielding ensures that radiation exposure remains within safe limits and is provided by the storage system. Radioactive decay of the spent fuel and the associated need for shielding will diminish as the storage period increases.

2.2.5. Retrievability and transportability

Retrievability is not necessary to maintain safety in the same sense as the previous four safety functions. Although safety can be achieved without maintaining retrievability, retrieval of the SFA or components of the storage system containing the spent fuel following storage could be necessary to enable subsequent stages of SFM. If future SFM stages require opening the package and handling SFAs, additional costs will be incurred if they are not readily retrievable. Hence, maintaining retrievability is important to the extent that it may minimize the costs and complexity of future SFM options.

Transportability, although not generally cited as a key safety function, also needs to be maintained throughout spent fuel storage to ensure that spent fuel can be moved to a reprocessing or disposal facility — or even to a new storage facility, with capabilities for inspection and repackaging if required. Due to the increased potential for materials degradation, maintaining retrievability and transportability may become more challenging over extended storage periods.

2.2.6. Active and passive systems

Safety functions can be performed by either active or passive systems. Passive systems do not rely on external inputs such as energy supply and mechanical actuators to perform their functions, although some human action may be needed periodically to confirm passive features remain functional (e.g. to verify no inlet/outlet coolant blockage). As spent fuel storage periods are extended, passive controls become increasingly important due to their increased reliability, lower operating cost, and reduced reliance on maintaining institutional controls [9].

Although the principles necessary for ensuring safe spent fuel storage currently and for extended periods are well understood, existing spent fuel storage systems may not have been designed for the time frames now contemplated, and safety analyses may not have fully considered the time frames now contemplated. This publication explores how these safety principles can be maintained when increasing operational lifetimes of existing spent fuel storage systems and how the effectiveness of future systems can be improved by including the possibility of extended storage in the design and regulatory frameworks of the future.

2.3. SUSTAINABILITY OF SPENT FUEL STORAGE

SSG-15 [10] defines short term storage as lasting up to approximately 50 years and long term storage beyond approximately 50 years and with a defined end point (reprocessing or disposal). SSG-15 [10] states that long term storage is not expected to last more than approximately 100 years, which is judged to be enough time to determine future fuel management steps.

The IAEA reports that because there are currently no operating spent fuel or HLW deep geological disposal facilities, spent fuel inventories in storage are growing, and much of the spent fuel will have to be stored for longer periods than initially intended, perhaps longer than 100 years [11]. The United States Nuclear Regulatory Commission considers 300 years of storage to be appropriate for the characterization and prediction of ageing effects and ageing management issues for extended storage and transport [7].

Continued generation and storage of spent fuel without full commitment to a clearly defined end point is not a sustainable policy. As stated in para. 1.6 of SSG-15 [10], "storage cannot be considered the ultimate solution for the management of spent fuel, which requires a defined end point such as reprocessing or disposal in order to ensure safety." Storage for longer and longer periods is not considered consistent with the responsibility to protect people and the environment without imposing undue burdens on future generations [2, 3]. Storage is by definition an interim measure that provides containment of spent fuel with the intention of retrieval for reprocessing, processing or disposal at a later time [4, 10, 12].

Some argue that deep geological disposal is the only generally accepted end point for spent fuel or for HLW from its reprocessing. Others argue that deep geological disposal could deny future generations the ability to utilize these materials for beneficial purposes. Given that spent fuel storage periods cannot be defined with certainty, maintaining SFM sustainability requires policies that ensure that continuing spent fuel storage will not impose an undue burden on future generations. This may be possible if, for all spent fuel inventories in storage, the financial, governance, technical and regulatory infrastructure for safely storing and achieving an acceptable end point is provided by the generations receiving the benefit.

3. AGEING MANAGEMENT PROGRAMMES

As laid out in IAEA Safety Standards Series No. SSG-48, Ageing Management and Development of a Programme for Long Term Operation of Nuclear Power Plants [13], ageing management programmes ensure safety is maintained through timely detection and control of age related degradation and other conditions that may jeopardize future handling and transport. Ageing management has been extensively applied in nuclear power plants as well as spent fuel storage facilities and addresses both the physical ageing and obsolescence of safety related SSCs [13–17]. Periodic safety assessments agreed with the regulator are designed to ensure the continued adequacy of power plant ageing management plans and to help to establish the technical basis for licence renewal. The same principles and approach apply to licence renewal for spent fuel storage facilities once the previous licence period approaches expiration. A properly executed ageing management programme will preclude unplanned or unanalysed conditions that could reduce safety or security margins or result in unnecessary remediation costs for longer storage periods. Information resulting from ageing management activities serves as a baseline for prioritizing needs and associated R&D objectives. It also provides context for the public and policy makers to understand properly the relative risks. Ageing management programmes also need to consider changes to policy and regulatory requirements relevant to longer storage periods.

3.1. AGEING MANAGEMENT FOR SPENT FUEL STORAGE

Evaluation of the consequences of the cumulative effects of both physical ageing and equipment obsolescence is an ongoing activity, illustrated by an adaptation of Deming's 'plan–do–check–act' cycle to the ageing management of SSCs, modified for a spent fuel storage facility (see Fig. 2) [13].

As stated in para. 2.21 of SSG-48 [13]:

"The closed loop of Fig. [2] indicates the continuation and improvement of ageing management on the basis of feedback on relevant operating experience, results from research and development, and results of self-assessment and peer reviews, to help to ensure that emerging ageing issues will be addressed."

Understanding the ageing of SSCs is key to effective ageing management and includes the following steps:

(a) Identification of SSCs: SSCs that are difficult to inspect, repair or replace are appropriately considered and identified in design and when planning monitoring and maintenance programmes.

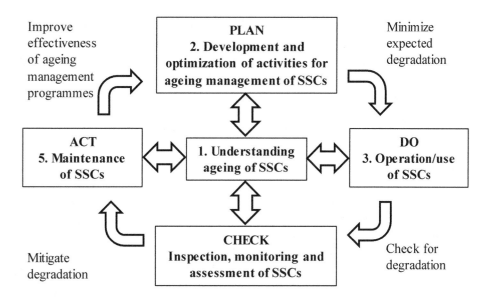

FIG. 2. Systematic approach to ageing management of SSCs for a spent fuel storage facility.

(b) Identification of applicable degradation mechanisms: Effective ageing management requires an understanding of the chemical, thermal, mechanical and radiation induced degradation processes and their combined effects. Research, development and testing may be required to understand consequences and identify methods to preclude, reduce or mitigate degradation of SSCs during extended operation of the spent fuel storage facilities.

(c) Consideration of external factors such as new technologies and regulatory changes relevant to spent fuel storage.

3.1.1. Plan

The 'plan' activity in Fig. 2 involves effective coordination and modification of existing programmes as well as the development of new programmes to ensure that prevention, detection and mitigation of ageing effects on SSCs is integrated into spent fuel storage facility operation, maintenance, monitoring and inspection plans, for example the plan to dry the cask cavity and boltholes before spent fuel loading to reduce corrosion during longer dry storage periods.

3.1.2. Do

The 'do' activity ensures proper function of SSCs by careful operation in accordance with operating procedures and technical specifications. Effective operations and maintenance programmes ensure that safety related SSCs are maintained, repaired or replaced as needed to prevent safety functions being compromised.

3.1.3. Check

The 'check' activity involves timely detection through effective monitoring and inspection programmes and assessment of any degradation of SSCs during extended storage, provided data are properly managed and preserved. Continuous measurement (monitoring) and periodic observation, measurement, testing and examinations (inspection) of SSCs are performed. Monitoring parameters and inspection intervals are established to ensure degradation is detected well before any loss of safety function or condition that may jeopardize future handling and transport. These parameters and intervals need to be periodically re-evaluated during longer storage times to consider historical trends, results of the past inspections and lessons learned from industry, among other things. Periodic reassessments of the condition of the spent fuel storage system with respect to evolving regulations and technology is also necessary to prevent obsolescence and to ensure compliance with the storage licensing basis throughout the storage period. Monitoring programmes consist of both condition and performance monitoring.

Condition monitoring searches for the presence and extent of ageing mechanisms that affect SSCs. These include measuring to determine the condition of pool structures, bedrock, concrete pads and other SSCs such as racks, protective coatings, cables and essential instrumentation. Some SSCs may be accessible to direct inspection only during service and others (e.g. components in high radiation fields) may require remote or other means of inspection and evaluation [18]. Generally, SFAs and spent fuel storage racks are more easily observed in wet storage than in dry storage.

Performance monitoring verifies the ability of the SSCs to perform their intended functions, for example:

(a) Shielding: Deviations in the trends between actual and calculated radiation levels may indicate shielding degradation.
(b) Containment: For metal storage casks, a check for leakage through pressure monitoring between cask lids can confirm the containment function of the metallic gasket as storage time lengthens. For concrete casks, the containment function can be checked by measuring the temperature difference between the top and bottom of the interior welded canister to confirm the presence of helium gas [19].
(c) Spent fuel integrity: In wet storage, monitoring and control of water chemistry will enable detection of spent fuel failure; whereas in dry storage a penetration into the canister or cask for cavity gas sampling may be required. Matsumara et al. [20] propose a non-destructive method to detect SFA failure by measuring gamma rays from ^{85}Kr released from the damaged SFA in the canister.
(d) Concrete structure in spent fuel storage facilities: The Schmidt hammer test can be deployed to detect any degradation.

3.1.4. Act

The 'act' activity involves the assessment of detected degradation (e.g. visual inspection, leak testing and radiation dose mapping), and the development and implementation of an appropriate corrective or mitigation plan to ensure no loss of safety function. Actions may include the following:

— Maintenance;
— Component repairs;
— Replacement;
— Design modifications;
— Updates to the operating, maintenance, monitoring and inspection plans.

Figure 2 utilizes feedback from relevant operating experience, results from self-assessment and R&D activities, and lessons learned from within the industry to ensure that emerging ageing management issues are identified and addressed. An example of the feedback loop in Fig. 2 can be found in the United States of America, where ageing processes that affect DCSSs have been compiled along with appropriate time limited ageing analyses (TLAAs) to assist licensees and regulators in understanding the issues and necessary actions associated with renewing spent fuel storage licences for extended storage periods [21]. The body of relevant knowledge to assist in developing the proper designs, operations and regulatory framework will continue to grow and evolve as spent fuel storage facilities age.

3.2. CLOSING DATA GAPS ON THE AGEING OF MATERIALS FOR SPENT FUEL STORAGE

Materials degradation needs to be well understood for both spent fuel and all system components relied upon to ensure confinement, shielding, decay heat removal and criticality safety and to maintain the ability for safe handling, retrieval and transport of the spent fuel package. For pool storage or dry storage vaults, this includes components such as pool liner materials and the facility structure. For DCSSs, the primary focus is to ensure the integrity of the canister or cask, including materials used in the cask body, trunnions, seals, neutron shielding and basket. Carlsen et al. [9] report (see also Refs [15, 22–37]):

"Though a broad knowledge base exists on the behaviour of existing SFAs and SFS facility materials, references for periods longer than a few decades are seldom found.... Further, evolving operating conditions and materials, such as higher burnup fuels and new fuel and cladding types, may also require testing to understand relevant age-related degradation under extended SFS.

....

"Several studies were recently conducted to identify possible degradation mechanisms, to analyse their potential impact on safety as SFS periods extend, and to identify knowledge gaps between anticipated technical needs and existing technical data....[2] Furthermore, a number of studies have identified the need for a full-scale HBU [high burnup] SF storage and transport confirmatory demonstration project as an essential part of developing the technical basis".

A number of national and international efforts are under way to address these knowledge gaps, including [9]:

— IAEA Coordinated Research Project on Demonstrating Performance of Spent Fuel and Related Storage System Components during Very Long Term Storage;
— Extended Storage Collaboration Program (ESCP), established by the Electric Power Research Institute;
— Used Fuel Disposition Campaign, established by the United States Department of Energy;
— Committee on the Safety of Nuclear Installations (CSNI) of the OECD Nuclear Energy Agency.

Closing these knowledge gaps will permit better modelling and analyses of ageing effects, which will enable efficient design and effective management and control of spent fuel storage facilities over longer periods. In the meantime, an awareness of these gaps enables continued safe spent fuel storage by allowing engineering design solutions to account for the associated uncertainties, for example knowledge gaps relating to cladding performance can be addressed by safety strategies and design approaches that reduce or avoid reliance on cladding integrity [38, 39].

4. DESIGN AND SITING OF FUTURE SPENT FUEL STORAGE SYSTEMS

Site selection and facility and equipment design can significantly reduce the risks and costs of spent fuel storage over extended periods. Currently, spent fuel is stored in: (i) water pools; (ii) welded metal canisters placed in concrete silos, casks or vaults; or (iii) bolted lid metal casks that provide shielding and other functions necessary to ensure safety. Both wet and dry storage technologies are used for either on-site at-reactor or away-from-reactor storage facilities as well as off-site storage facilities (see Ref. [40] for a summary).

Much of the focus on extending storage has been on developing the technical basis for ensuring that existing SFA and packaging components will continue to perform their credited safety functions during extended storage. Although this is a necessary activity for extending the storage periods of existing facilities, recognition and acceptance that spent fuel can be stored for multiple licensing periods provides opportunities to include upfront design and functional requirements that will improve the effectiveness of future spent fuel storage facilities.

The inventory of spent fuel presently in storage is the result of only 50 years of a relatively small yet growing industry. Given the present and likely future role of nuclear power, the current spent fuel inventory will be only a small fraction of the future inventory that will need to be stored until sufficient reprocessing or disposal capacity becomes available. Hence, most of the spent fuel storage facilities that will be needed have not yet been designed or built.

---

[2] Identified gaps were prioritized based on the associated risk [33, 34].

Future spent fuel storage facilities can benefit by considering designs that facilitate extending storage durations and that can adapt to different safety strategies that may be necessary as a result of changing conditions, regulations and societal values over expanded storage periods. Although these design considerations may result in increased upfront investment, the life cycle costs may be lower than those incurred by more traditional approaches that presume static conditions over the spent fuel storage facility lifetimes. Facility design, site selection and spent fuel storage configurations can significantly impact the issues and associated costs for extending storage.

4.1. SYSTEM DESIGN BASIS

Spent fuel storage systems designed to accommodate prolonged storage will need to contemplate a broader range of scenarios that could occur over the longer time period. These include the potential for increased magnitude and likelihood of challenges due to natural phenomena such as floods and earthquakes, the accrued effects of ageing; and the impacts of changing societal values and policies [9]. For designs to accommodate uncertain spent fuel storage periods, it is thus appropriate to consider increasing safety margins to accommodate the potentially broader range of conditions that may be encountered during longer storage time. Designs for extended spent fuel storage need to include provisions to mitigate age related stressors for safety related SSCs. Packaging and drying technologies ought to be selected to reduce the likelihood of future cladding failures. Fuel storage and package configurations ought to facilitate inspection and monitoring systems. Phenomena such as mechanical, thermal, chemical, radiation and other stresses that can accumulate or change over time need to be considered. Simple and robust system design as well as appropriate materials selection and quality controls are considered essential for the long term reliability of radiological barriers and other systems relied upon for safety and security. Structural materials need to be chosen and environmental conditions controlled to provide adequate resistance to corrosion, stress corrosion cracking and other age related degradation. Design and operational controls that reduce the frequency and magnitude of operational transients, such as temperature cycles and mechanical handling, will further reduce fatigue and other challenges to structural materials. Designs for lengthened spent fuel storage also need to consider the challenges associated with maintaining quality controls and records necessary for longer storage periods.

In addition to ensuring the longevity and facilitating the maintenance of safety related SSCs, the design of spent fuel storage facilities also needs to consider the importance of maintaining public confidence. Hence, grounds, paint and other visible indicators that the facility is being properly maintained are also important. For similar reasons, future designs may consider features to facilitate interaction with the public and to provide other community benefits.[3]

Spent fuel storage facility and equipment designs for extended storage consider the possibility that spent fuel in storage may require remediation to ensure safe transport and compatibility with future SFM steps. Hence, spent fuel storage facilities should be able to maintain, confirm and, if needed, restore transportability. Considerations such as measures to facilitate retrievability and standardize handling equipment become increasingly important as the storage duration increases. For example, placing one or more SFAs into a standardized basket, prior to storage, could support a number of objectives. The use of such a basket could:

— Provide features to standardize handling of various spent fuel;
— Provide features to facilitate monitoring, inspection and proper data management and preservation;
— Provide a means for assuring that spent fuel can be retrieved for inspection or repackaging without reliance on maintaining the structural integrity of the SFA;
— Enhance criticality safety by providing additional structural support for geometry control and a means for adding neutron poisons or for displacing moderators;
— Allow spent fuel to be stored in a variety of configurations (e.g. casks, pools and vaults) and to be readily transferred from one configuration to another if needed.

---

[3] The HLW inside the HABOG facility, the Netherlands, will gradually decay until future generations and governments decide on the method of disposal of the radioactive waste. This process of decay is symbolized by the orange colour of the building, selected by its designer, Ewoud Verhoef, because it is halfway between red and green. The exterior of the building will be periodically repainted in successively lighter shades until it reaches white in about 100 years, by which time the thermal output of the waste will have reduced by one order of magnitude.

In theory, spent fuel storage facilities could be designed with materials, inspection and maintenance capabilities to support operations for perhaps several hundred years. However, when the associated costs are considered, it may be preferable in some instances to design for a more modest lifetime and purposely plan for significant refurbishment of facilities and equipment and possible repackaging of spent fuel at that time. The methodology and means for eventual decommissioning are also key considerations in designing new spent fuel storage facilities and extending existing facilities.

4.1.1. Ageing management and licence renewal considerations

New spent fuel storage systems that accommodate uncertain storage periods not only integrate lessons from past ageing management into their design basis but also consider design features to facilitate future ageing management plans. These features can include the provision for monitoring and inspection systems as well as the means to repair or replace key components. Examples of refurbishment that could be necessary during increased storage periods include repair of concrete structures (pools or casks), repacking spent fuel into new storage canisters or replacing neutron absorber panels in spent fuel pools.

Monitoring and inspection systems can better support extended storage by considering advanced surveillance and non-destructive examination techniques to monitor storage conditions and to support ageing management through both preventive and predictive maintenance that provide for the following:

— Access to key components for periodical surveillance;
— Accurate baseline and in-service inspection records, inclusion of material test specimens and other features that facilitate unambiguous interpretation of inspection results;
— Detection of operating conditions that could challenge materials performance or accelerate degradation;
— Early detection and accurate analyses of degradation affecting safety related SSCs.

In light of the potential time frames for extended storage, spent fuel storage facilities may consider additional infrastructure such as hot cells, pools and remote handling equipment to allow for maintenance, repair and refurbishment or replacement of key components when needed. Capabilities may include the means to accomplish the following:

— Remediate, repackage or relocate spent fuel or packaging components that have degraded during storage or cannot be verified to meet applicable requirements;
— Restore transportability if the spent fuel package cannot be demonstrated to meet requirements following extended storage (i.e. spent fuel package conditions or transport requirements could change);
— Management of radioactive waste streams.

By developing and incorporating design features that facilitate ageing management, conducting materials performance research and relicensing activities needed to ensure continued safety, the costs and risks associated with extending spent fuel storage may be considerably reduced.

4.1.2. Safeguards and security considerations

Facilities for extended spent fuel storage will need to fulfil current and future requirements prescribed by relevant conventions and treaties on safeguards and physical protection. From a safeguards standpoint, it is very important to have the ability to verify the nuclear material stored and to keep continuity of knowledge of the nuclear material. It is also important to have implemented safeguards by design [41, 42] and to comply with the national regulations. Suitable measures need to be applied to prevent unauthorized access to, or removal of, radioactive material. As such, safety and security considerations need to be an integral part of facility design and siting.

Many of the technical considerations important to safety also apply to safeguards and security. Spent fuel storage technologies are designed to maintain their safety functions under severe natural events. These attributes also provide similar protections against diversions or hypothetical attacks. However, additional barriers and monitoring systems will eventually be needed to meet safeguards and security requirements under extended storage due to the decreased radiation fields as spent fuel ages. This radiation is a significant factor in determining the security features to guard against theft and diversion. Although a reduced radiation field lowers the safety risk, it also diminishes the self-protecting characteristic[4] of the spent fuel, which may result in the need for additional physical barriers, monitoring and personnel requirements to protect the fuel.

In addition to the eventual loss of the self-protecting characteristic, security requirements can vary significantly over time. The security needed to protect spent fuel is based in part on the perceived threat, which can change significantly as a result of external events. Consequently, the development and periodic review of rational and realistic threat scenarios will function as a tool to assess the security measures needed to ensure the proper protection of spent fuel during extended storage.

4.2. WET VERSUS DRY STORAGE FACILITIES

Following discharge from the reactor, spent fuel is typically stored in cooling pools for a minimum of 3–5 years. To make space for newly discharged spent fuel or when the reactor facilities are eventually decommissioned, this spent fuel is eventually removed from the cooling pool and transferred to pools, dry storage vaults or DCSSs.

Water filled pools for storing spent fuel have proven to operate reliably and safely for several decades. Decay heat is removed by the water where it is transferred to cooling water through heat exchanges or to ambient air. By controlling water temperature and chemistry, the potential for degradation and consequent release of radioactive inventory is kept low. In addition, the significant thermal inertia of large pool water volumes provides considerable grace periods for taking remedial actions in case of abnormal plant conditions or accidents. Potential disadvantages of spent fuel pools include increased reliance on active safety controls (e.g. maintaining water levels, water chemistry, cooling and make-up systems, and leak detection) and possible additional degradation mechanisms caused by being in a wet environment. Wet storage of spent fuel requires considerably more radioactive waste management operations (e.g. liquid waste treatment) and generates a waste stream that will require additional waste processing.

Dry spent fuel storage systems includes DCSSs designed for a single canister or cask and large vaults designed to store many SFAs. Dry spent fuel storage systems are typically designed to rely on passive safety features. Decay heat is transferred to the surface of the storage containment system through conduction and radiation, where it is transferred to ambient air by natural convection. Dry spent fuel storage facilities have also been shown to operate reliably over several decades. Potential disadvantages of dry spent fuel storage systems include the following:

— Higher fuel temperatures and associated heat load limitations;
— More complex equipment and radiological protection measures if handling bare fuel assemblies in a dry environment;
— Lack of direct access to the spent fuel for inspection.

Wet storage pools and most dry vault storage facilities typically store spent fuel as bare assemblies, whereas with DCSSs spent fuel is typically placed into sealed canisters or casks prior to storage (see Section 5). DCSSs can be readily deployed when reactor storage pools reach capacity and added incrementally as needed. This 'pay as needed' approach can reduce the upfront capital investment needed for spent fuel storage. Due to increased reliance on passive features for safety and security, the operational costs may also be lower, along with transport costs, because the need for handling bare spent fuel at shipping and receiving facilities is avoided. Relative to current DCSS designs, the construction of spent fuel storage pool or dry vault storage facilities requires a larger capital commitment and may also result in increased operational expense associated with active systems such as water treatment or heating, ventilation and air-conditioning systems. This additional expense may, however, be offset by the larger capacity of pool and dry vault facilities that capitalize on the economies of scale.

---

[4] The International Panel on Fissile Materials report that [43]: "For about the first 100 years, LWR spent fuel emits gamma radiation at a dose rate greater than 1 sievert per hour, which would be lethal to about 50% of adults (LD50) in three to four hours. At such exposure, the IAEA considers irradiated spent fuel sufficiently radioactive that it could only be moved and processed with specialized equipment and facilities, beyond the practical capabilities of sub-national groups, and therefore 'self protecting'."

4.3. CENTRALIZED SPENT FUEL STORAGE FACILITIES

Current spent fuel storage facilities are typically co-located within the reactor sites. This could result in the need for on-site (at-reactor or away-from-reactor) spent fuel storage facilities to be maintained long after the associated reactor has been decommissioned if reprocessing or disposal of the spent fuel is not accomplished before the end of the reactor's licensed service life. A consequence could be either increased risks due to reductions in on-site operational infrastructure or increased storage operational costs to the extent that this infrastructure is maintained solely to support continued spent fuel storage. In addition, it could also reduce the incentive to decommission the reactor in a timely manner because the spent fuel storage facility is still on-site — thereby leading to greater challenges and higher risks, along with additional challenges of maintaining records and knowledge preservation. Furthermore, it is likely that the affected communities would have neither considered, nor consented to, these reactor sites becoming long term spent fuel storage facilities.

Storing spent fuel in one or more regional spent fuel storage facilities provides several benefits. In addition to enabling reactor sites to remove significant radioactive hazards and fully decommission at the end of reactor lifetime, centralized spent fuel storage facilities may also significantly reduce the costs of operations, maintenance and security by avoiding the need to duplicate these same costs across many sites. For example, centralized spent fuel storage reduces the licensing burden by allowing the consolidation of information, expertise, equipment and other infrastructure for:

— Information management systems and other institutional controls;
— Inspection, monitoring and analyses of spent fuel and its packaging component;
— Repair, refurbishment or replacement of safety related or other key SSCs;
— Repackaging or other equipment needed to respond to contingencies or other events;
— Any preparation or pre-staging steps needed prior to disposal or reprocessing.

The risks associated with obsolescence of equipment are also reduced by standardizing and maintaining equipment at consolidated spent fuel storage facilities. Carlsen et al. [9] find "in particular, if transport of aged SF packages should come into question because the infrastructure for restoring transportability (i.e., inspection, repackaging, overpacking, etc.) could be very costly if duplicated at multiple sites." Furthermore, the economies of scale will also allow consideration of pool, vault and other storage alternatives that may not be cost effective for multiple smaller facilities. Similarly, the cost and complexity of future transport and handling may be considerably reduced due to standardization of equipment and handling methods at centralized facilities.

A potential obstacle to centralized spent fuel storage is the difficulty of finding a site that satisfies technical, societal and political criteria. It might be especially difficult to garner public acceptance by a host community if they receive no benefit from the reactor operations generating the spent fuel or have concerns about the site becoming a de facto repository [9]. Other obstacles include: (i) potentially larger consequences from severe events that could affect the entire population of stored fuel; (ii) additional costs and risks (albeit small) of transport to the facility; and (iii) the upfront investment costs of centralized storage, which may be greater than the incremental storage costs at existing sites for short timescales, but likely offset by economic of scales when total life cycle costs are compared and contingencies for the risks of repackaging are included.

4.4. SITING CONSIDERATIONS

Most of the facilities needed to store spent fuel until sufficient reprocessing or disposal capacity becomes available have not yet been designed, licensed or built. This presents an opportunity to reduce substantially many of the risk and cost factors associated with extending spent fuel storage by prudent site selection for new spent fuel storage facilities:

— Corrosion and other challenges to materials performance can be significantly reduced by building the facility in a dry, temperate climate.
— Natural hazards can be significantly reduced by selecting a location with a low probability of severe events (e.g. floods and earthquakes).

— Human made hazards can be significantly reduced by selecting sites in areas that are isolated from industrial or other potential hazards and easily protected from hostile actions.
— Potential environmental and safety impacts of an unforeseen event can be significantly reduced by building the facility where the consequences would be low (e.g. small populations, and limited or already restricted water and land use).
— Because both wet and dry facilities transfer decay heat to the environment, the availability of cooling water and ambient air temperatures are considerations.
— Transport routes and equipment compatibility between spent fuel storage, reprocessing and disposal facilities are more easily managed for co-located facilities and, if co-located with a disposal facility, the need for transport is eliminated.

IAEA Safety Standards Series No. NS-R-3 (Rev. 1), Site Evaluation for Nuclear Installations [44], contains criteria and methods that can be used in a graded approach in the siting of spent fuel storage facilities.

5. SPENT FUEL STORAGE CONFIGURATIONS

Spent fuel storage configurations can accommodate uncertain storage periods, to facilitate ageing management and to provide flexibility for future steps to achieve an acceptable end point. Key decisions in selecting a spent fuel storage configuration to meet present and future needs relate to how spent fuel will be stored, whether spent fuel will be packaged for transport or disposal prior to or after storage, what components will be relied upon to perform essential safety functions, and how safety performance will be demonstrated with sufficient certainty to satisfy regulatory requirements. Each decision affects future options. Available alternatives are evaluated to select a strategy that can be sustained over extended storage periods while maintaining flexibility and adaptability to accommodate the full range of plausible future scenarios.

Geological repositories will require spent fuel to be placed in suitable disposal containers before emplacement underground. For many States pursuing repositories, the design (e.g. capacity and material specifications) for the disposal container and the acceptance criteria for the contained waste form are not yet settled. This has significant implications for spent fuel storage on how and when spent fuel is placed into containers.

5.1. CANISTERS AND CASKS

Placing spent fuel into robust canisters or casks prior to storage provides a number of benefits. The spent fuel cladding, often relied upon to confine radiological materials and to maintain the geometry needed for shielding and criticality safety, is difficult to inspect and not possible to repair or replace. The canister or cask serves as an additional barrier "because it ensures confinement of radionuclides, enhances criticality safety by precluding intrusion of a moderator and maintain an inert environment that precludes oxygen, humid air, and water, which could initiate or accelerate degradation processes" [9]. The spent fuel canister or cask also provides a means of retrievability and handling in the event that the structural integrity of the fuel is compromised. In short, the canister or cask can serve as an inspectable, repairable and/or replaceable component that can perform key safety functions — thus reducing the need to rely on the integrity of the spent fuel and its cladding for assuring safety over extended storage periods [9]. Other considerations that favour storing spent fuel in canisters or casks are that the spent fuel is in a configuration that can be more readily relocated and the canister may also provide increased robustness relative to severe accident conditions.

However, for spent fuel that is sealed in storage canisters or casks, there is limited monitoring capability and no access for direct inspection of the spent fuel or canister internals. Knowledge of the condition and structural integrity of spent fuel and its packaging is an essential part of the present safety basis for storage and handling and also for demonstrating compliance with post-storage requirements for transport and acceptance for reprocessing or disposal. Having little direct access for inspection will necessitate additional costs associated with the development of a design and licensing basis to accommodate the additional uncertainty relating to the aged condition of the spent fuel and its packaging. Inaccessibility of the spent fuel and canister or cask internals for regular inspection may also result in missed opportunities for early detection and mitigation of unforeseen degradation mechanisms.

Without the possibility for direct inspection, a more comprehensive, predictive capability of degradation mechanisms is needed. This may increase the research, development and demonstration work needed to develop the technical and regulatory basis for demonstrating that spent fuel and its packaging will meet applicable requirements for storage and transport following storage. Furthermore, the size and weight of the additional packaging materials could limit transport and handling options. Future acceptance criteria for reprocessing and disposal facilities can also place limitations on the form and composition of materials. For these reasons, it may become necessary to open and repackage spent fuel that has been packaged before storage. A successful packaging strategy that avoids the need for future repackaging requires a canister or cask suitable for the duration of storage, subsequent transport and eventual disposal (if that is the end point). This approach would benefit from:

— Successful ageing management programmes;
— Selection of canister or cask properties (i.e. geometries, heat loading, criticality safety measures, and materials) compatible with foreseeable repository acceptance criteria;
— Stability of applicable policies and regulations;
— Accomplishing reprocessing or disposal before repackaging becomes necessary.

If spent fuel is packaged prior to storage, a more robust strategy might be to assume that repackaging will eventually be necessary and to design packaging and operational strategies to facilitate it. An approach that plans for periodic repackaging can:

— Provide a basis for cost planning;
— Enable periodic inspection to confirm compliance with performance requirements and to obtain data to support R&D needs as well as underpin predictive spent fuel performance capability;
— Allow for renewal and updating of SFA and packaging components and monitoring equipment to capitalize on new technologies and to address new or changed requirements.

However, repackaging adds risks, costs and personnel exposure and also generates radioactive waste. Cost estimates for future repackaging greatly depend on discount rates and other assumptions used in the economic analysis. Some estimates have shown that periodic repackaging could increase costs by an order of magnitude or more. The United States Government Accountability Office, reporting on the termination of the Yucca Mountain Repository programme, estimates that the cost of DCSSs could increase from US $30–60 million to US $180–500 million per reactor based on an assumption of repackaging operations about every 100 years [45].

Although the costs and impacts of repackaging could be reduced by simply overpacking existing packages, decision makers should be cautious about repackaging approaches that simply place the package into successive overpacks. Experience has shown that this can add complexity and uncertainty to future operations. Each additional packaging component makes inspections more difficult, increases the size and weight of the package and reduces heat transfer. These changes can adversely affect the package performance for operational states and accident conditions.

5.2. BARE ASSEMBLIES

The alternative to packaging spent fuel prior to storage is to store it as bare assemblies. Bare spent fuel can be stored in pools or dry vaults that provide shielding and other necessary safety features. A key advantage of storing bare, unpackaged spent fuel is the increased access for monitoring and inspection throughout the storage period. Thus, any degradation that might occur over extended storage is more readily detected. This greatly facilitates demonstrating compliance with safeguards requirements during storage and with transport requirements following longer storage periods. It provides additional information to support the design of handling and transport equipment and reduces uncertainties and the associated need to broaden design and safety margins. Furthermore, because heat generation and radiation decrease during storage, the need for shielding and heat removal decreases over time, which can allow a reduction in the size, weight and cost of future packaging.

Another advantage of postponing packaging until after the storage period is that there will be much less uncertainty relative to the requirements for transport and acceptance criteria for reprocessing or disposal — thus minimizing the likelihood that repackaging will be necessary. Postponing packaging will also allow future packaging designs to capitalize on future technologies and materials and will provide a 'fresh' package for post-storage transport and handling. This could substantially reduce the technical gaps and associated R&D needed to predict the condition of spent fuel stored in sealed containers. Thus, an engineering approach that relies on canisters or individual cans rather than spent fuel cladding integrity may lessen the burden on cask designers and regulators to do extensive research on spent fuel cladding properties [39]. Lastly, storing bare spent fuel that can be packaged for transport and disposal after storage allows repository design and selection to proceed without being constrained or influenced by decisions relating to spent fuel packaging.

Because the potential for change in both the condition of the spent fuel and its packaging components, as well as in regulatory requirements, increases over time, it is also reasonable to assume that the likelihood of needing to repackage or take other mitigation measures can be positively correlated with the interval between when the spent fuel was last packaged and when it is transported. Hence, compliance with future requirements is more readily achieved with strategies that postpone packaging or that include a ready means for inspecting the spent fuel and packaging components and for remediating or repackaging when needed.

6. REGULATORY CONSIDERATIONS

By considering the potential for multiple licence renewals, regulatory frameworks can be designed to ensure safety over the storage periods that may be necessary until an acceptable end point is achieved.

6.1. ROLE OF REGULATORS

The regulator provides oversight of SFM activities until an acceptable end point is achieved. Regulatory review and oversight provides a nexus in ensuring safety and maintaining active controls by [9]:

(a) Periodically renewing, or reassessing the safety basis, of storage facilities and associated ageing management programmes;
(b) Approving new technologies;
(c) Enhancing current regulatory frameworks with new approaches (e.g. risk informed guidance) to address uncertainties while maintaining safety during prolonged storage;
(d) Ensuring appropriate compatibility and integration with reprocessing and disposal regulatory frameworks (current and future).

Key regulatory roles may include (but not be limited to) the following [9]: (i) research activities to inform regulatory frameworks, develop licensing basis and support decisions; (ii) enhancement and implementation of licensing, inspection and oversight activities; and (iii) public outreach and communication. It is recognized that, depending on the State, regulator responsibilities may not encompass each of these items. For example, some regulatory authorities might not have independent research responsibilities. They could instead be responsible for making informed decisions based on available research. Depending on the scheme, the regulator may also have additional responsibilities relating to increasing spent fuel storage periods.

6.1.1. Regulatory research activities

Ageing management and storage concepts need to be considered holistically by regulators, industry and research institutions. Regulatory research activities may include experiments, technical studies and analyses:

— To assess the safety significance of potential technical issues;
— To develop the long term licensing basis;
— To improve the long term endurance of spent fuel storage systems.

Research needs to be designed for evaluating new ageing management and design approaches as well as for evaluating and verifying ageing management programmes (e.g. periodic reassessments of facilities). This includes research needed to implement and verify time limited ageing analyses (see Section 3.1), as well as monitoring, maintenance and mitigation activities to ensure safety over the licensing period. In order to identify research needs, regulators, industry and research institutions perform regulatory gap studies to identify priorities. It is important to identify and implement long lead research items that may require early planning to address needs several years in the future.

6.1.2. Licensing, inspection and oversight

The role of regulators is to ensure that regulations, guidance and inspection programmes are adequate to guide development and to assess implementation of ageing management programmes that will ensure safety and maintain the ability to store and transport over each successive licensing period. Regulators also ensure that operational experience is collected during storage and rules are updated to address emerging issues.

The primary focus of the licence renewal review, or periodic reassessment, will be the validity of the time limited ageing analyses and adequacy of the ageing management programme proposed for each specific storage design (i.e. including the frequency at which major cask components need to be inspected or renewed). A key element of the regulatory inspection programme will be verifying the actual implementation of each ageing management programme and appropriate corrective actions by the licensee. In addition, regulators may establish operational experience programmes to identify generic technical issues and ageing trends as well as other programmes or requirements to assure that records, knowledge management and other infrastructure are maintained to ensure continued safe storage and to demonstrate compliance with future transport, reprocessing or disposal criteria. It is also appropriate that regulators reconsider both the likelihood and severity of design basis accident conditions based upon the possibility of multiple renewals that may significantly increase storage periods [9].

As part of licence renewal, the regulator ensures that adequate financial resources remain available to meet financial obligations. Spent fuel storage costs include operations, security personnel, monitoring, equipment and facility maintenance and replacement, possible inspection and repackaging operations and eventual decommissioning. These ongoing operating costs will increase as spent fuel inventories grow and spent fuel storage equipment and facilities age. Consequently, the financial liability and the associated need for financial assurances may grow significantly as storage periods are extended.

Each State will have different laws and financial structures for achieving financial assurance. Given the uncertainties in storage time frames, regulators need to consider a variety of plausible scenarios and periodically re-evaluate and adjust scenarios, as needed, to ensure financial resources will remain available to meet these obligations for the duration of spent fuel storage.

6.1.3. Public outreach

Public confidence is key to ensuring sustainable SFM and consequently sustainable nuclear power. Public outreach in a regulatory context will be an essential element of storing spent fuel for the unknown periods that may be required until an end point is achieved. By proactively reaching out, the industry can engage interested parties to explain extended storage activities and the basis for the safety, security and protection of people and the environment. The regulator can focus on increasing understanding of the actual risks and impacts of storage activities and enhancing confidence in its role for enforcing safety. The regulator also impacts public acceptance by effectively communicating ongoing activities and seeking feedback (as appropriate under law) on policy changes and regulations that affect extending storage.

6.2. REGULATORY FRAMEWORK FOR EXTENDING STORAGE

The regulatory framework consists of the national policies, rules, guidance and technical standards established to ensure the continued safety, security and environmental protection of spent fuel storage. Policy makers and regulators periodically reassess to ensure the regulatory framework appropriately reflects technology advances, evolving societal values and needs, industry needs and other emerging issues.

Because spent fuel discharges will exceed disposal and reprocessing capacity for the foreseeable future, the means for safely extending the spent fuel storage period will thus be an implicit part of an effective SFM programme. The foreseeable significant increases in spent fuel inventories, combined with the uncertainty in the spent fuel storage periods, present an opportune time to re-examine the approaches and assumptions that underpin spent fuel storage licensing processes. Carlsen et al. [9] find:

"Key licensing considerations may include the frequency for relicensing, lead-times for initiating relicensing, the appropriate length of time that should be assumed for ageing performance analyses, and the frequency of key ageing management actions."

Licence expiration does not represent a limit on the duration that spent fuel can be stored safely. Licence periods are normally determined by regulatory or policy considerations and typically represent an interval considered sufficiently long to meet the anticipated spent fuel storage period. Figure 3 illustrates how a regulatory framework can be designed to include periodic licence renewal to address the present situation (i.e. unknown storage periods) while ensuring safe storage until a national policy for a spent fuel end state is decided and implemented. The licence renewal process is represented by the upper loops in the figure. Carlsen et al. [9] report:

"The success of this approach depends upon a licensing renewal process that can reliably identify any vulnerability that could jeopardize successful SFS over the length of the next licensing period and on the ability of the licensee to take effective corrective actions if and when needed to qualify for license extension. License applications would need to be submitted well in advance of license expiration to provide sufficient time for any needed corrective actions. Corrective actions could range from enhanced monitoring or inspection programs to remediation of degraded SF containers or storage facilities, repackaging of spent fuel assemblies (SFA), or even relocating to a new facility."

If the approach illustrated in Fig. 3 is effectively implemented, the number of licence renewals need not be limited because compliance with requirements will be assessed and confirmed one finite step at a time (i.e. licensing period).

6.3. RISK INFORMED PERFORMANCE BASED REGULATION

Regulators might consider the advantages of risk informed, performance based approaches for addressing the technical uncertainties of long term storage [9]. Some regulatory frameworks are currently based on deterministic approaches where performance objectives (e.g. dose limits) and subsidiary design requirements must be satisfied for specified sets of operational states and accident conditions. Ageing management programmes are also generally deterministic or prescriptive in nature, based on conservative engineering judgment.

Based on the definition from the Nuclear Regulatory Commission [46]:

"A 'risk-informed' approach to regulatory decision-making represents a philosophy whereby risk insights are considered together with other factors to establish requirements that better focus licensee and regulatory attention on design and operational issues commensurate with their importance to health and safety."

The US regulation 10 CFR 63, Disposal of High-level Radioactive Wastes in a Geologic Repository Yucca Mountain, Nevada [47], incorporates many of these principles (e.g. dose limits for off-normal and accident conditions during pre-closure operations were correlated with the likelihood of the scenario). Similar principles could be applied to regulate the risks associated with extending spent fuel storage periods.

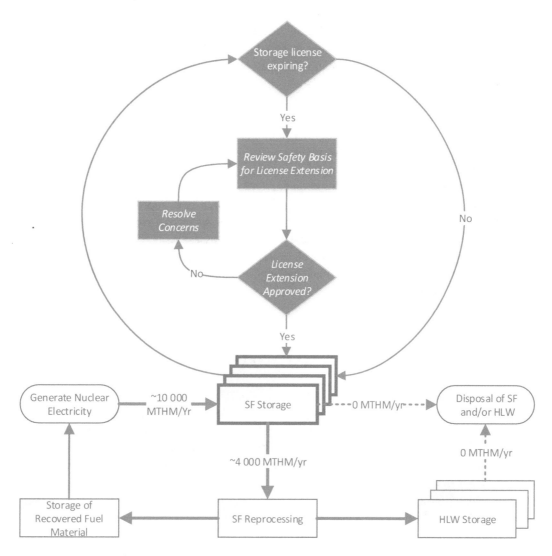

FIG. 3. Licence renewal until reprocessing or disposal is accomplished.

Risk informed approaches may also enhance the traditional ageing management and design requirements by allowing explicit consideration of a broader set of potential safety challenges and by providing a logical means for prioritizing challenges based on risk significance, operating experience and engineering judgement. Development of risk informed analyses will also help to provide a better understanding of the level of protection provided by traditional deterministic approaches.

Risk analyses provide a better understanding of the probability and consequence of specific ageing degradation failures, which may offer insights and alternative approaches for ensuring safety. For example, some spent fuel storage safety functions are often allocated to spent fuel cladding, which helps to confine radiological materials and maintain the geometry of the SFA. Cladding integrity is difficult to inspect and not feasible to repair or replace. For extended storage periods, it might be prudent to consider safety strategies that shift safety functions to spent fuel packaging components or spent fuel facility features that can be more readily monitored and inspected and, if needed, repaired or replaced. This is similar to the current practice of 'canning' to provide an additional barrier and means of handling for suspect SFAs. Other solutions for reducing risks associated with reliance on cladding integrity include use of packaging and drying technologies to safeguard against future cladding failures and of designs to facilitate inspection and monitoring systems to assess the internal spent fuel storage environment and fuel cladding conditions periodically.

Performance based approaches establish requirements based on satisfying specified performance criteria without explicitly prescribing the methods for meeting the criteria. Given the uncertainties associated with the possibility of multiple spent fuel storage licence extensions, risk informed and performance based approaches may encourage development of new technologies and more effective approaches for ensuring long term safety [9].

Regulatory and operational frameworks sometimes include implicit assumptions about: the storage duration; the condition of the fuel, packaging and equipment technologies employed at time of transport; or the subsequent stages of SFM management. These implicit assumptions may influence or needlessly limit potential solutions.

A performance based regulatory approach is also independent of assumptions about what the future may hold. It focuses on assuring safe conditions while leaving flexibility to the licensee as to the means of meeting established safety criteria. Hence, performance based regulation can provide the flexibility to accommodate evolving technologies and policies. In providing this flexibility, the regulator is also encouraged to consider the licensee's need for clearly defined and objective requirements.

Enhancing regulatory storage, transport and disposal frameworks with both a risk informed and performance based approach focuses attention on the most important activities and establishes objective criteria based upon risk insights for evaluating performance. Key challenges in implementing new approaches for storage and transport in this manner include: (i) defining the level of acceptable risk for spent fuel storage; (ii) providing the upfront investment of resources for data collection and analysis needed to assess risks; (iii) establishing appropriate risk metrics and performance requirements that provide clear and objective guidance for licensees; and (iv) maintaining an appropriate level of defence in depth. Additional challenges include the integration of risk informed frameworks across the entire back end of the fuel cycle.

7. POLICY CONSIDERATIONS

Sustainable SFM requires policies and strategies to provide a clear, consistent and stable direction because they drive the need for spent fuel storage as well as the available options and timing for achieving an acceptable end point such as reprocessing or disposal. Until States address spent fuel reprocessing or disposal on a sufficient scale to accommodate their spent fuel discharges, storage for ever longer periods will be necessary. This is not considered consistent with the responsibility to protect human health and the environment without imposing undue burdens on future generations [2, 3].

The development of SFM policy is a complex process involving politics, economics, resource conservation, environmental protection and public perception, the last of which has become a predominant factor in many States [48]. Policy decisions heavily influence the trajectory of nuclear energy and establish the context within which the entire nuclear fuel cycle is managed.

Within the growing worldwide energy demand, projections of nuclear energy demand vary considerably, largely owing to policy related variables. National energy strategies and policies affect preferences, subsidies and mandates, which in turn affect the economic landscape for energy decisions. Nuclear energy demand is generally projected to be much higher in models where policies favour a reduction in carbon emissions or do not favour specific classes of energy [49].

Accounting for variations and considering the effects of policy changes in the aftermath of the events at Fukushima, nuclear generation capacity is estimated to grow by 35–100% by 2030 — from 370 GW(e) in 2010 to 500–750 GW(e) [11]. In addition, environmental and societal pressure for reduced greenhouse gas emissions and potential new markets for nuclear power (e.g. hydrogen production and water purification) represent significant future demands.

Since spent fuel discharged will increase proportionally with nuclear power production[5], sound policies that encourage effective long term SFM decisions are crucial. Policy decisions directly impact the quantity of spent fuel and the times that storage will be needed. They also influence technical considerations, such as storage technologies and design approaches, and have a significant impact on overall trust in the nuclear fuel cycle. It is important to recognize the interconnectivity between SFM issues and to consider them holistically when developing policies affecting SFM.

---

[5] Next generation nuclear plants driven by fast reactors can reduce the spent fuel discharged per unit of energy produced by factors of around 100 as well as provide other inherent safety and security benefits. However, the development and implementation of these technologies on a commercial scale will require decades. A clear and credible policy on the development and use of these reactor technologies will be necessary to attract the investment and sustained commitment to develop, license and implement advanced nuclear fuel cycles.

7.1. A DEFINED END POINT

National policy determines the end point for spent fuel — whether it be reprocessing or disposal. Similarly, key decisions such as on-site versus off-site centralized or regionalized spent fuel storage facilities are also primarily a matter of national policy. Hence, clear policies, objectives and strategies are essential for the safety and sustainability of SFM, and for the adequate allocation of financial and human resources over time frames that span many generations [50].

A clearly defined and credible end point for spent fuel is needed for effective decisions and planning. Although spent fuel storage serves several beneficial purposes in the back end of the fuel cycle (e.g. radioactivity and heat reduction, and time for development of technologies), these benefits can be fully realized only with a well defined end point and strategy for achieving it. Without a defined end point as an objective, all which remains is a wait-and-see or find-and-fix approach to SFM. Either could prove very costly in the event that changes to storage configurations or facilities become necessary, for example as a result of unforeseen degradation of the spent fuel or packaging components or changes in spent fuel policies or requirements.

7.2. CLEAR OWNERSHIP AND ACCOUNTABILITY FOR SPENT FUEL

National policies establish ownership, accountability and liability for entities responsible for each portion of the nuclear fuel cycle and for each activity within the scope of SFM. Different entities often have responsibilities for decisions affecting the characteristics of spent fuel and its storage configuration and conditions. This includes fuel design (e.g. fuel and cladding material selection, and fabrication techniques), reactor operations (e.g. burnup) and SFM (e.g. package configuration and storage conditions).

Each entity tends to make decisions that optimize performance relative to its responsibilities. Thus, fuel designers and reactor operators are likely to optimize fuel designs and operational parameters for the most effective power production. This tendency toward local optimization might limit storage and disposal options or needlessly add to costs and complexity.

Policies that establish ownership, accountability and liability affect the time horizon and decision considerations of SFM. As noted in Section 5, SFM decisions on when and how spent fuel is packaged can significantly impact costs and options for future storage, transport and disposal. There are several available strategies, and the relative costs and benefits of each depend on the length of spent fuel storage and who pays for it, who pays for packaging and, if needed, any repackaging, and who pays for disposal. Furthermore, different storage and packaging options are available if storage can be centralized. Each of these considerations is primarily a matter of policy.

As discussed in Section 8.1, it is important to design and implement spent fuel policies that encourage all participants in the nuclear fuel cycle to take a holistic approach and make decisions that properly consider other portions of the fuel cycle as well as the impact on effectively reaching an end point. Since the government will eventually assume responsibility for long term stewardship, policies should consider the benefit of establishing the infrastructure to facilitate centralized management and custody at the appropriate stage of SFM.

7.3. SUSTAINABLE SPENT FUEL MANAGEMENT

As noted in Section 2.2, extending spent fuel storage need not be viewed as passing an undue burden to future generations if the means for assuring an acceptable end point are also passed along. This would include the necessary financial resources, governance and regulatory infrastructure, technical capabilities, and records and information, among other things [10].

For States with small nuclear programmes that may require many decades of nuclear electricity production to accrue the capital necessary for implementing an end point, this approach may be attractive or even necessary. The Netherlands, for example, has designed and constructed a facility intended to store HLW for at least 100 years — with the objective of completing the development work and accruing the means during this 100 year period to enable a future generation to decide to either implement the end point, continue storage (with appropriate renewal and relicensing) or consider other alternatives that may then be available.

This approach may be viewed as empowering future generations to make their own decisions with respect to continued storage, beneficial reuse or implementation of an end point in the context of the perceived risks, societal needs and values, and future, available technologies. Giving the choice to future generations, however, comes with some risks. Provision of adequate infrastructure and financial reserves assures sustainability only in a stable economic and political environment. Furthermore, States can misjudge the resources required or their ability to preserve them. Future generations could lose or misuse the resources, thus depriving other future generations of the same opportunity. These are valid concerns that policy makers will need to balance. If, however, spent fuel is considered as a valuable resource, there is little risk that spent fuel storage will become a burden to future generations.

7.4. MULTINATIONAL APPROACHES

Several advantages have been associated with consolidating spent fuel storage (see Section 4.3) into one or more regional facilities. Multinational spent fuel storage facilities that accept fuel from several countries may provide similar benefits on larger scales. The economics of a multinational spent fuel storage facility may be especially favourable for countries in close proximity to one another or with relatively small nuclear programmes. However, such a facility may pose additional challenges in addressing the unique values of multiple countries and navigating more complex political, legal and financial challenges.

7.5. STABLE LONG TERM POLICIES

The siting, design and licensing of nuclear facilities has proven to be a decades long process with relatively high uncertainty of success, due largely to challenges relating to public acceptance and changes in political support that occur over these time frames. Experience has shown that policies based on attempts to predict or legislate on future outcomes have not fared well.

Stable policy for SFM has been difficult to achieve in many States. Societal perceptions of nuclear power have evolved from unbridled optimism in the 1950s to unfounded fear in the 1980s and back to cautious optimism. The cycles have varied somewhat from country to country and have been influenced by nuclear accidents such as those at Three Mile Island, Chernobyl and Fukushima, as well as by the cost and availability of other energy sources. Attitudes toward nuclear power remain strongly influenced by public opinion and other factors that are not conducive to the long term stability of nuclear policy. Hence, effective policies for SFM are designed to be insulated from short term or event driven political reactions or swings in public opinion while still being responsive to longer term changes in underlying societal values.[6]

Recent projections of the growth of nuclear power have been fuelled largely by growing energy demand owing to the increasing consensus on the need to reduce carbon and other greenhouse gas emissions — coupled with recognition of the limitations on the availability of fossil fuels and on the costs and scalability of renewables. Weighing the long term risks of nuclear power against the long term risks of an energy future without it appears to provide a perspective that is favourable for the growth in nuclear energy production. It is important that this longer term perspective be nurtured and incorporated into SFM policies.

Past attempts have presumed a foreseeable future for the management of spent fuel and its by-products. However, significant costs and delays were often encountered due to the inability to account for the effects of unforeseen events and policy changes. Recognizing unpredictable future politics, many States are now working to develop phased, adaptive strategies that move forward toward an end point in slow, measured steps. This approach provides time for the public and policy makers to become familiar with the issues, technologies, costs and benefits, and to work out mutually acceptable agreements before committing. Stable, long term policies might best be achieved by thoughtful inclusion of provisions that allow a path to slowly evolve and adapt as the future unfolds.

---

[6] For example, owing to increased concerns relative to terrorist threats and also to the accident at the Fukushima Daiichi nuclear power plant, policy makers in some countries presumed that wet storage involved higher risks, and introduced policies that may have needlessly limited storage options or resulted in additional costs and personnel exposure without commensurate safety benefits [51].

SFM is a long term proposition. Conditions, information, available resources and technologies, politics and societal values change over time. Hence, periodic assessments of policies and their effects on SFM are necessary to ensure that they encourage an objective, thorough and holistic approach to SFM and that satisfactory progress is being made toward an end point. Effective assessments will consider realistic time frames and scenarios for implementing reprocessing or disposal of spent fuel along with the associated environmental, financial and societal risks and trade-offs. Finally, such assessments provide confidence to society that its concerns are being heard and appropriately addressed.

7.6. RELIANCE ON NEAR TERM SOLUTIONS

Developing sufficient spent fuel reprocessing or disposal capacity to achieve an end point for spent fuel requires substantial capital investment over several decades. Moreover, repository siting and licensing pose significant technical challenges along with many social and political obstacles. On the other hand, spent fuel storage technologies have proven to be safe and relatively economical. As a result, it has been difficult to garner the political will to commit the financial and political capital necessary to achieve an acceptable end point. Nonetheless, the costs and risks associated with maintaining spent fuel storage will continue to escalate as the spent fuel inventory and storage periods increase and, if not addressed, will eventually become a significant societal burden.

The accumulation of spent fuel in storage can be viewed as a symptom of an issue for which there are two available paths — a short term fix and a fundamental solution. The fundamental solution is to provide sufficient reprocessing and disposal capacity. This is capital intensive and requires a sustained, long term commitment before the benefit is realized. Hence, continued reliance on the short term fix of simply adding storage capacity and extending storage durations is very attractive. However, although this symptomatic solution provides immediate relief, it tends to further weaken the political will and, over time, also the societal resources needed to achieve the fundamental solution.

In system dynamics, this well known pattern is commonly referred to as 'shifting the burden'.[7] Although it is recognized that symptomatic solutions are sometimes needed as a transitional solution to stabilize the system or postpone until the fundamental solution can be applied, it is nonetheless crucial to recognize and manage this behavioural pattern, as it tends to lead to a repeated application of a symptomatic solution that is not sustainable while undermining the ability to achieve a lasting solution.

In the case of SFM, there is a professional and ethical obligation to ensure safe and effective spent fuel storage for as long as necessary. However, development of the capability to extend storage periods safely is not to be used as an excuse to delay implementation of the processes needed to achieve an acceptable end point. Delaying the fundamental solution will result in escalating inventories and SFM costs, thus making it progressively more difficult to manage. Depending on the public perception of the risks associated with a deep geological repository relative to those associated with continued spent fuel storage, delays in reaching an end point and growing spent fuel inventories could also negatively impact the public confidence needed to move forward with a lasting solution. Therefore, a strong caution is given that policies relying on continuing spent fuel storage, though presently necessary, should be managed to ensure commitment to achieving a sustainable SFM policy.

8. OTHER KEY CONSIDERATIONS

8.1. MANAGING INTERFACES THROUGHOUT THE FUEL CYCLE

The nuclear fuel cycle comprises three major phases: the front end, reactor operations and the back end. These phases are divided into closely interrelated discrete stages (e.g. storage, transport, and reprocessing or disposal for the back end). Decisions, considerations and available options in each phase may not only be constrained by those taken in previous phases, but may also affect those of subsequent phases. In recognition of this, Art. 5 of Council

---

[7] See www.systems-thinking.org/theWay/ssb/sb.htm

Directive 2011/70/Euratom of 19 July 2011 establishing a Community framework for the responsible and safe management of spent fuel and radioactive waste [52] states:

"1. Member States shall establish and maintain a national legislative, regulatory and organisational framework ('national framework') for spent fuel and radioactive waste management that allocates responsibility and provides for coordination between relevant competent bodies."

Article 4(3) [52] stipulates that "National policies shall be based on the of the following principles: ... (b) the interdependencies between all steps in spent fuel and radioactive waste generation and management shall be taken into account". Some regulatory frameworks or different safety approaches or practices in the storage, transport and reprocessing or disposal of spent fuel could result in compatibility issues that could increase costs or risks as spent fuel storage periods are extended. For example, design, operational and regulatory decisions may not necessarily consider scenarios in which spent fuel is stored for extended periods in sealed canisters. This may result in the need to inspect packaging and SFAs, and possibly for repackaging in order to meet further storage, transport or disposal requirements.

A key challenge in achieving sound decisions with respect to safe, secure, effective and sustainable SFM is that overall responsibility, accountability and ownership for all stages, including reprocessing or disposal of spent fuel, need to be clear. Dividing this responsibility among the different phases of the fuel cycle without sufficient interaction between interested parties fosters localized decisions that may limit future options and increase life cycle costs. A holistic approach to the management of the entire fuel cycle becomes increasingly important as spent fuel storage periods are extended. It enables decisions that consider the entire life cycle and discourages local optimization achieved at the expense of overall system performance.

Integration across the design, operational, regulatory and policy aspects of fuel cycle management enables decisions that ensure the compatibility of materials, operations, equipment, packaging and waste form with future SFM steps, thus reducing risks and life cycle costs. For example, decisions relating to the expected timing and acceptance requirements of a geological repository will impact the relative advantages and disadvantages of various storage configurations. These storage technology choices, in turn, significantly impact the costs and impacts of extending storage. Storage technology decisions may also influence the design criteria, options and costs for future repository design and siting. This interdependence within the nuclear fuel cycle applies across all phases of the nuclear fuel cycle. Examples of the effects to phases of the fuel cycle include the following:

(a) Fuel fabrication and reactor core management and operations: Utilities manage operating costs while maintaining high reliability and safety standards. Short term cost considerations favour fuel designs and operating parameters that allow higher burnup and minimize fuel failures. This encourages continued development and use of new designs and advanced materials for fuel, cladding and assembly structures as well as the reactor coolant chemistry. Changes to these parameters can affect the structural and corrosion properties of spent fuel and cladding materials and thus affect available options and costs for extended storage and subsequent transport.

(b) Spent fuel storage: The interdependency between storage and subsequent transport for reprocessing or disposal strongly influences decisions on how and when spent fuel is placed into packages. For example, there are two fundamentally different approaches to the design and siting of a geological disposal facility. One approach seeks the optimum disposal system, unconstrained by any existing packaging. The other seeks a disposal system that accommodates existing spent fuel packages. Clearly, the entity responsible for the repository would prefer the former while the entity responsible for storage would favour the latter (i.e. allowing fuel to be packaged optimally for storage objectives). Hence, if one entity (e.g. the utility) is responsible for spent fuel storage decisions while another (e.g. the government) is responsible for transport or disposal, then SFM decisions during storage are unlikely to fully consider their impacts beyond spent fuel storage and may limit or significantly impact the future costs of transport and disposal.

(c) Spent fuel transport: Transport licences may expire while fuel is in storage or requirements for storage licences and transport licences may not be fully compatible. Hence, both transport and storage licences are to be maintained for spent fuel in storage, which is a requirement in some States. In Japan, for example, a holistic approach was established with a 50 year storage and transport licence. The same principle could be applied as storage periods are extended.

(d) Spent fuel reprocessing: Reprocessing operations typically utilize undamaged fuel to reduce contamination when unloading transport packages. If receiving spent fuel that may have degraded during storage, reprocessing facility design and operational changes may be needed to manage potential contamination. In addition, significant degradation during storage could affect reprocessing if it becomes difficult or costly to accurately characterize the spent fuel.

(e) Spent fuel disposal: Current designs of mined deep geological repository typically utilize disposal packages with smaller thermal, weight and fissile material capacities than many dual purpose casks in use. Yet the trend is toward casks that reduce fabrication and operational costs by maximizing payload (i.e. increasing size, fissile and thermal limits). In addition, the material and structural requirements needed to meet repository acceptance criteria are not yet specified. Hence, it is possible that spent fuel currently being stored may need to be repackaged prior to disposal.[8] The need to accept and repackage existing canisters or to dispose of existing canister designs may place significant additional siting and design constraints on the operator of a future deep geological repository.

8.2. PUBLIC CONFIDENCE

Public confidence in SFM is a key factor in siting and licensing storage and disposal facilities. Factors affecting public confidence have been extensively studied and a number of resources to aid in understanding and addressing the topic are available (see Ref. [53]). Nonetheless, achieving societal acceptance and the associated political wavering has proven to be a greater challenge than maintaining safety and security. Seaborn [54] finds:

"the public concerns and uneasiness related to nuclear fuel waste and nuclear power in general...derive, among other things:

- from the association with nuclear weapons and past disasters like Chernobyl;
- from the mystery which for most people surrounds nuclear fission and the longevity of radiation;
- from the fear of disastrous consequences if 'something would go wrong', however unlikely that 'something' is in statistical terms;
- from the uneasiness about a waste management system which does not envisage indefinite monitoring of what is happening in a disposal vault and the surrounding geosphere;
- from a lack of confidence in the ability of scientists to predict what is likely to happen ten thousand years from now;
- from difficulties in determining how best to protect the interests of future generations when we make decisions now about the nuclear wastes we have generated;
- from a feeling that there must be some better and less dangerous way to generate the electrical power we need.

"Some of this apprehension and skepticism can be refuted at one level by scientific arguments. But the concerns nevertheless remain and are real. They have to be given serious attention in developing public policy in a democracy."

8.2.1. Public confidence dilemma

Public confidence with regard to the management of spent fuel and HLW is heavily influenced by perceived risks as well as a lack of confidence in, or understanding of, SFM. Much has been invested in sharing information and in assuring and demonstrating safety in order to address directly these concerns — with limited success relative to building public confidence.

---

[8] It has been suggested that packaging into smaller canisters or casks would provide an increased likelihood of compatibility with future storage, transport and disposal configurations. Use of smaller packages provides more flexibility, but also adds costs and operational impacts without the assurance that this investment will preclude the need for future repackaging.

Failure to achieve sufficient public support has been a persistent source of difficulties, delays and challenges to maintaining the political needed for successful siting and licensing of a deep geological repository for disposal of spent fuel or HLW. Missed commitments and continued difficulty in making substantive progress reinforce risk perceptions and further erode public confidence that spent fuel and HLW can be sustainably managed, which increases both the quantity of spent fuel and the duration of spent fuel storage while further increasing the challenges associated with siting and licensing spent fuel storage facilities. This circular effect is illustrated in Fig. 4. Ironically, a lack of public confidence and political will to address SFM can go full circle and ultimately aggravate the situation further, impeding a lasting solution. However, public confidence is influenced not only by perceived risk but also by perceived benefits, suggesting that both areas present opportunities for bolstering public confidence.

8.2.2. Perceived risks versus perceived benefits

A key observation from Fig. 4 is that public confidence is actually part of a balance that considers both risks and rewards. Hence, although adequate safety performance is necessary, reducing risks and risk perceptions alone is unlikely to be sufficient to reverse the cycle. This will require that benefits also be recognized and valued.[9] Building public confidence requires that the public recognize the opportunity for real benefit, or for avoiding real cost, relative to their values. Without this incentive, there is no motive for taking a position.

Although public confidence has proven to be relatively resistant to past efforts to defend or improve the nuclear safety record, it has also proven to be receptive to the benefits of nuclear power, with public support increasing in many countries.[10] This is true even after considering the effects of the tsunami and subsequent damage to the Fukushima Daiichi nuclear power plant. The recent changes in the public's perception of nuclear energy are largely attributed to recognition of the benefits of nuclear power relative to other large scale energy producing technologies (e.g. very low life cycle carbon emissions and increased energy security).

Spent fuel can and must be managed safely. However, public confidence is determined not only by perceived risk but also by perceived benefits. Experience shows that an excellent safety record is not sufficient to gain public confidence. Benefits commensurate with the perceived risks also have to be recognized. To build the public confidence necessary for effective siting and operation of spent fuel storage facilities, the benefits of nuclear power also need to be recognized and valued by society. The communities that provide the necessary services need to benefit accordingly. Effective siting processes capitalize on the benefits, so the communities selected to host SFM facilities consider themselves as winners rather than losers.

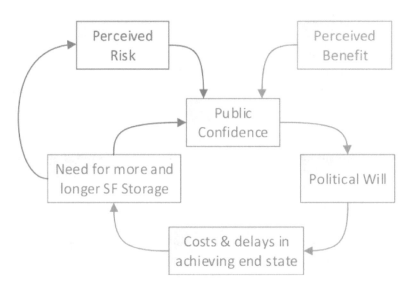

FIG. 4. Public confidence and feedback.

---

[9] Perceived risks and benefits are influenced by many factors, for example, risk perception is affected by the level of understanding of nuclear phenomena, trust in nuclear authorities and the nuclear safety record, among other things. The perception of benefit from nuclear energy is influenced by energy needs and costs.

[10] See www.foratom.org/publications/#topical_publications

Benefits of nuclear power include reliability, low carbon emissions, stable long term fuel supply, geographic flexibility, and an excellent environmental and safety record. Benefits to spent fuel storage host communities that can persist long afterwards include economic strength from increased employment, infrastructure and other incentives associated with hosting the facility. However, realizing these benefits and earning the trust needed for progress will require continued efforts to better understand and more effectively communicate relevant issues, the development of benefits that are stable and resilient to uncertain economic and political conditions, and the political will to make the long term investments and commitments needed to stay the course.

8.2.3. Public acceptance and extending spent fuel storage

The decision to generate and use nuclear power represents a commitment by society to the long term stewardship (i.e. safe storage, reprocessing or disposal) of the spent fuel that is generated. Even in States that have elected to phase out nuclear power, spent fuel and HLW exist from past use; and significant quantities of spent fuel continue to be generated worldwide.

Spent fuel storage is an essential part of the nuclear fuel cycle and extending spent fuel storage can provide several benefits with respect to overall SFM. The additional cooling time will reduce heat, radiation and the associated safety and design constraints. It is also beneficial for the development and implementation of SFM policies and the associated technologies. Despite these potential benefits, extending spent fuel storage may also be viewed as a first step on a path toward continued deferral of decisions, thus rendering the storage facility a de facto disposal location. This viewpoint cannot be easily countered by public officials when there are no reliable plans for a deep geological repository, or if a country has already faltered on past promises to remove spent fuel.

The same factors that produce the need for continuing storage (i.e. no clearly defined end point or delays in achieving it) also create challenges with respect to achieving public acceptance both for siting new spent fuel storage facilities and for extending the licences of existing spent fuel storage facilities. If spent fuel is not proactively managed in view of a clear and credible end point such as reprocessing or disposal, continued spent fuel storage is likely to be viewed as the result of an inadequate or failed policy. The need for, and benefits of, extending spent fuel storage are unlikely to be accepted if viewed as an undesirable but necessary interim step resulting from a deferral of decisions or a lack of alternatives.

Although many States struggle to site deep geological repositories, there are common threads among States that have succeeded in achieving a higher measure of public and political support. These lessons can also be applied to siting new spent fuel storage facilities and relicensing existing ones. A commonality among those that are moving forward successfully is that the siting process was not pushed or imposed on the public, but rather was chosen by them through an open dialogue addressing both risks and benefits (see Ref. [55]). Other factors include:

(a) Early, honest and ongoing communication by the organizations responsible for SFM as well as the regulators to address public concerns and distrust.
(b) A stepwise implementation whereby the local public gains knowledge and familiarity with the risks, and both understands and begins to derive benefit before fully committing.

A focus on communities that already host nuclear facilities. They have the most at stake and are more familiar with nuclear power and its associated risks. Experience worldwide has shown that siting progresses quickest among these communities.

After an extensive review of international repository development programmes, the Blue Ribbon Commission, United States of America, "recommends a siting process that is consent-based, transparent, phased, adaptive, standards- and science-based, and governed by legally-binding agreements between the federal government and host jurisdictions" [8].

8.3. INSTITUTIONAL CONTROLS

Unlike a repository, spent fuel storage, by definition, will remain under institutional control until an acceptable end point is achieved. Applicable institutional controls include the full spectrum of regulatory requirements discussed in Section 6. As noted in Section 6.2, a regulatory framework based on periodic renewal need not presume how long the storage period might last. Each renewal independently confirms that the organizational, financial, technical and material requirements needed to ensure safety throughout the licensing period are satisfied. This is somewhat analogous to the way in which airplanes and vehicles are licensed. They are not forcibly retired based on age but are allowed to remain in service as long as the applicable requirements are met. In addition, the periodic relicensing process provides an opportunity for institutional controls to evolve with societal needs and values and also to consider new technologies and options that may become available.

A potential concern associated with reliance on periodic relicensing to evaluate and confirm that spent fuel storage will continue to meet applicable requirements is that, if requirements are not satisfied, withholding a storage licence will not eliminate the risk. If the licensee, for any reason, fails to satisfactorily maintain the licence, the responsibility for enforcing requirements and ensuring safety will ultimately fall on the government. Hence, the need for clearly defined responsibilities and liabilities, and the associated financial assurance, take on increased importance. There are a number of societal analogues where institutional control has persisted over centuries. Prisons, education, health care, arts, religions and cultural traditions have persisted through many generations.[11] Society attends to matters considered important, so it is expected that spent fuel will remain under institutional control for as long as it is considered to be a hazard. Spent fuel will remain hazardous for many centuries, so the question of institutional control becomes largely an issue of proper application of the principles of ethics and sustainability — specifically of ensuring that the burden is not to be passed to future generations. With respect to extending spent fuel storage, proper consideration of institutional control is largely a matter of assuring the financial, human and technical resources necessary for safe and effective storage and disposal or reprocessing of the spent fuel and disposal of any associated waste. Hence, the status and infrastructure for maintaining the adequacy of these resources need to be periodically evaluated in the relicensing process for extending spent fuel storage periods.

On account of the longevity of the hazard, the unknown storage duration and the uncertainty with regard to subsequent stages of SFM, the preservation of records, knowledge and memory is of particular interest. The OECD Nuclear Energy Agency has initiated a multidisciplinary activity aimed at increasing collaboration and identifying best practices with regard to the development and implementation of plans for the preservation of records, knowledge and memory.[12] Although focused primarily on processes relating to the development and management of a geological repository, many of the principles and lessons learned can be applied to extended spent fuel storage.

---

[11] Examples include: the Hudson Bay Company, established in 1670 (Canada); continuous management of underground quarries for safety in Paris since the 1400s (France); and the Shikinen Sengu of Ise Jingu shrine, which has been rebuilt every 20 years since the 700s (Japan).

[12] See www.oecd-nea.org/rwm/rkm

REFERENCES

[1] INTERNATIONAL ATOMIC ENERGY AGENCY, Nuclear Fuel Cycle Objectives, IAEA Nuclear Energy Series No. NF-O, IAEA, Vienna (2013).

[2] EUROPEAN ATOMIC ENERGY COMMUNITY, FOOD AND AGRICULTURE ORGANIZATION OF THE UNITED NATIONS, INTERNATIONAL ATOMIC ENERGY AGENCY, INTERNATIONAL LABOUR ORGANIZATION, INTERNATIONAL MARITIME ORGANIZATION, OECD NUCLEAR ENERGY AGENCY, PAN AMERICAN HEALTH ORGANIZATION, UNITED NATIONS ENVIRONMENT PROGRAMME, WORLD HEALTH ORGANIZATION, Fundamental Safety Principles, IAEA Safety Standards Series No. SF-1, IAEA, Vienna (2006).

[3] OECD NUCLEAR ENERGY AGENCY, The Roles of Storage in the Management of Long-lived Radioactive Waste: Practices and Potentialities in OECD Countries, OECD, Paris (2006).

[4] INTERNATIONAL ATOMIC ENERGY AGENCY, IAEA Safety Glossary: Terminology Used in Nuclear Safety and Radiation Protection (2007 Edition), IAEA, Vienna (2007).

[5] INTERNATIONAL ATOMIC ENERGY AGENCY, Nuclear Technology Review 2015, IAEA, Vienna (2015).

[6] ELECTRIC POWER RESEARCH INSTITUTE, Nuclear Sector Roadmaps: Used Fuel and High-Level Waste Management, EPRI, Palo Alto, CA (2013).

[7] NUCLEAR REGULATORY COMMISSION, Plan for the Long-term Update to the Waste Confidence Rule and Integration with the Extended Storage and Transportation Initiative, Rep. SECY-11-0029, Office of Nuclear Material Safety and Safeguards, Washington, DC (2011).

[8] UNITED STATES DEPARTMENT OF ENERGY, Strategy for the Management and Disposal of Used Nuclear Fuel and High-Level Radioactive Waste, USDOE, Washington, DC (2013).

[9] CARLSEN, B., et al., "Challenges Associated with Extending Spent Fuel Storage until Reprocessing or Disposal", Safety of Long-term Interim Storage Facilities (Proc. Workshop, Munich, 2013), OECD, Paris (2013) 43–60.

[10] INTERNATIONAL ATOMIC ENERGY AGENCY, Storage of Spent Nuclear Fuel, IAEA Safety Standards Series No. SSG-15, IAEA, Vienna (2012).

[11] International Status and Prospects for Nuclear Power 2012: Report by the Director General, GOV/INF/2012/12-GC(56)/INF/6, IAEA, Vienna (2012).

[12] Joint Convention on the Safety of Spent Fuel Management and on the Safety of Radioactive Waste Management, INFCIRC/546, IAEA, Vienna (1997).

[13] INTERNATIONAL ATOMIC ENERGY AGENCY, Ageing Management and Development of a Programme for Long Term Operation of Nuclear Power Plants, IAEA Safety Standards Series No. SSG-48, IAEA, Vienna (2018).

[14] NUCLEAR REGULATORY COMMISSION, Standard Review Plan for Review of License Renewal Applications for Nuclear Power Plants, Rep. NUREG-1800, Rev. 2, Office of Nuclear Reactor Regulation, Washington, DC (2010).

[15] NUCLEAR REGULATORY COMMISSION, Generic Aging Lessons Learned (GALL) Report, Rep. NUREG-1801, Rev. 2, Office of Nuclear Reactor Regulation, Washington, DC (2010).

[16] INTERNATIONAL ATOMIC ENERGY AGENCY, Plant Life Management for Long Term Operation of Light Water Reactors: Principles and Guidelines, Technical Reports Series No. 448, IAEA, Vienna (2006).

[17] INTERNATIONAL ATOMIC ENERGY AGENCY, Plant Life Management Models for Long Term Operation of Nuclear Power Plants, IAEA Nuclear Energy Series No. NP-T-3.18, IAEA, Vienna (2015).

[18] INTERNATIONAL ATOMIC ENERGY AGENCY, Understanding and Managing Ageing of Material in Spent Fuel Storage Facilities, Technical Reports Series No. 443, IAEA, Vienna (2006).

[19] TAKEDA, H., WATARU, M., SHIRAI, K., SAEGUSA, T., Development of the detecting method of helium gas leak from canister, Nucl. Eng. Des. **238** (2008) 1220–1226.

[20] MATSUMURA, T., SASAHARA, A., NAUCHI, Y., SAEGUSA, T., Development of monitoring technique for the confirmation of spent fuel integrity during storage, Nucl. Eng. Des. **238** (2008) 1260–1263.

[21] ARGONNE NATIONAL LABORATORY, Managing Aging Effects on Dry Cask Storage Systems for Extended Long-term Storage and Transportation of Used Fuel, Rev. 0, ANL (2012).

[22] AMERICAN SOCIETY FOR TESTING AND MATERIALS, Standard Guide for Evaluation of Materials Used in Extended Service of Interim Spent Nuclear Fuel Dry Storage Systems, ASTM C1562-10(2018), ASTM, West Conshohocken, PA (2010).

[23] AMERICAN SOCIETY FOR TESTING AND MATERIALS, Standard Practice for Evaluation of the Long-Term Behavior of Materials Used in Engineered Barrier Systems (EBS) for Geological Disposal of High-level Radioactive Waste, ASTM C1174-17, ASTM, West Conshohocken, PA (2017).

[24] NUCLEAR REGULATORY COMMISSION, Standard Review Plan for Renewal of Spent Fuel Dry Cask Storage System Licenses and Certificates of Compliance, Rep. NUREG-1927, Office of Nuclear Material Safety and Safeguards, Washington, DC (2011).

[25] NUCLEAR REGULATORY COMMISSION, Dry Cask Storage Characterization Project — Phase 1: CASTOR V/21 Cask Opening and Examination, Rep. NUREG/CR-6745, Office of Nuclear Regulatory Research, Washington, DC (2001).

[26] NUCLEAR REGULATORY COMMISSION, Examination of Spent PWR Fuel Rods after 15 Years in Dry Storage, Rep. NUREG/CR-6831, Office of Nuclear Regulatory Research, Washington, DC (2003).

[27] ELECTRIC POWER RESEARCH INSTITUTE, Dry Cask Storage Characterization Project, Rep. 1002882, EPRI, Palo Alto, CA (2002).

[28] HANSON, B., et al., Gap Analysis to Support Extended Storage of Used Nuclear Fuel, Rev. 0 (2012).

[29] UNITED STATES NUCLEAR WASTE TECHNICAL REVIEW BOARD, Evaluation of the Technical Basis for Extended Dry Storage and Transportation of Used Nuclear Fuel, NWTRB, Arlington, VA (2010).

[30] NUCLEAR REGULATORY COMMISSION, Materials Aging Issues and Aging Management for Extended Storage and Transportation of Spent Nuclear Fuel, Rep. NUREG/CR-7116, Office of Nuclear Material Safety and Safeguards, Washington, DC (2011).

[31] ELECTRIC POWER RESEARCH INSTITUTE, Extended Storage Collaboration Program (ESCP): Progress Report and Review of Gap Analyses, Technical Report 1022914, EPRI, Palo Alto, CA (2011).

[32] ELECTRIC POWER RESEARCH INSTITUTE, Extended Storage Collaboration Program: International Subcommittee Report — International Perspectives on Technical Data Gaps Associated with Extended Storage and Transportation of Used Nuclear Fuel, Technical Report 1026481, EPRI, Palo Alto, CA (2012).

[33] NUCLEAR REGULATORY COMMISSION, Identification and Prioritization of the Technical Information Needs Affecting Potential Regulation of Extended Storage and Transportation of Spent Nuclear Fuel, NRC, Washington, DC (2012).

[34] STOCKMAN, C.T., HANSON, B.D., ALSAED, A.A., "Used nuclear fuel storage and transportation data gap prioritization", Integrating Storage, Transportation, and Disposal (Proc. 14th Int. High-level Radioactive Waste Management Conf. Albuquerque, 2013), Vol. 1, Curran Associates, Red Hook, NY (2013) 503–508.

[35] BLUE RIBBON COMMISSION ON AMERICA'S NUCLEAR FUTURE, Report to the Secretary of Energy, United States Department of Energy, Washington, DC (2012).

[36] XIHUA HE, et al., Development and Evaluation of Cask Demonstration Programs (2011).

[37] NICHOL, M., et al., "Concept plan for a high burn-up fuel storage and transportation confirmatory data project", Integrating Storage, Transportation, and Disposal (Proc. 14th Int. High-level Radioactive Waste Management Conf. Albuquerque, 2013), Vol. 1, Curran Associates, Red Hook, NY (2013) 94–103.

[38] WEAVER, D., "Future licensing strategies for the storage and transportation of spent fuel", presentation at the 27th Spent Fuel Management Sem. Arlington, 2012.

[39] EASTON, E.P., BAJWA, C.S., ZHIAN LI, GORDON, M., "Licensing strategies for the future transportation of high burnup spent nuclear fuel", presentation at the ASME 2012 Pressure Vessels and Piping Division Conf. Toronto, 2012.

[40] INTERNATIONAL ATOMIC ENERGY AGENCY, Survey of Wet and Dry Spent Fuel Storage, IAEA-TECDOC-1100, IAEA, Vienna (1999).

[41] INTERNATIONAL ATOMIC ENERGY AGENCY, Facility Design and Plant Operations Features that Facilitate the Implementation of IAEA Safeguards, STR-360, IAEA, Vienna (2009).

[42] INTERNATIONAL ATOMIC ENERGY AGENCY, International Safeguards in Nuclear Facility Design and Construction, IAEA Nuclear Energy Series No. NP-T-2.8, IAEA, Vienna (2013).

[43] INTERNATIONAL PANEL ON FISSILE MATERIALS, Managing Spent Fuel from Nuclear Power Reactors: Experience and Lessons from Around the World, IPFM (2011).

[44] INTERNATIONAL ATOMIC ENERGY AGENCY, Site Evaluation for Nuclear Installations, IAEA Safety Standards Series No. NS-R-3 (Rev. 1), IAEA, Vienna (2016).

[45] UNITED STATES GOVERNMENT ACCOUNTABILITY OFFICE, Commercial Nuclear Waste: Effects of a Termination of the Yucca Mountain Repository Program and Lessons Learned, GAO-11-229, GAO, Washington, DC (2011).

[46] NUCLEAR REGULATORY COMMISSION, White Paper on Risk-informed and Performance-based Regulation, SECY-98-144, NRC, Washington, DC (1999).

[47] NUCLEAR REGULATORY COMMISSION, Disposal of High-level Radioactive Wastes in a Geologic Repository at Yucca Mountain, Nevada, 10 CFR 63.

[48] INTERNATIONAL ATOMIC ENERGY AGENCY, Status and Trends in Spent Fuel Reprocessing, IAEA-TECDOC-1467, IAEA, Vienna (2005).

[49] SON H. KIM, WADA, K., KUROSAWA, A., ROBERTS, M., Nuclear energy response in the EMF27 study, Clim. Change **123** (2014) 443–460.

[50] INTERNATIONAL ATOMIC ENERGY AGENCY, Policies and Strategies for Radioactive Waste Management, IAEA Nuclear Energy Series No. NW-G-1.1, IAEA, Vienna (2009).

[51] "Risks and benefits of accelerated fuel transfer examined", Nucl. News **55** (2012) 38–41.

[52] Council Directive 2011/70/Euratom of 19 July 2011 establishing a Community framework for the responsible and safe management of spent fuel and radioactive waste, Official Journal of the European Union No. L 199, Publications Office of the European Union, Luxembourg (2011).

[53] OECD NUCLEAR ENERGY AGENCY, Stakeholder Confidence in Radioactive Waste Management: An Annotated Glossary of Key Terms, OECD, Paris (2013).

[54] SEABORN, B., "Criteria for public support for a waste management concept: The Environmental Assessment Panel's recommendations", Public Confidence in the Management of Radioactive Waste: The Canadian Context (Workshop Proc. Ottawa, 2002), OECD, Paris (2003) 71–74.

[55] OECD NUCLEAR ENERGY AGENCY SECRETARIAT, "International perspective", Ibid., pp. 31–33.

BIBLIOGRAPHY

Ageing management programmes

BENDERESKAYA, O.S., et al., "Corrosion of research reactor Al-clad spent fuel in water, reduced enrichment for research and test reactors" (Proc. Int. Mtg San Carlos de Bariloche, 2002).

BERNSTEIN, R., GILLEN K.T., Fluorosilicone and Silicone O-ring Aging Study, Sandia Report SAND2007-6781, Sandia National Laboratories, Albuquerque, NM (2007).

BILLINGTON, D.S., CRAWFORD, J.H., Radiation Damage in Solids, Princeton University Press, Princeton, NJ (1961).

BOLTON, G., "Innovative technologies for condition monitoring of waste containers", paper presented at Nuclear Interim Storage 2011, London, 2011.

CALVERT CLIFFS NUCLEAR POWER PLANT, Calvert Cliffs Independent Spent Fuel Storage Installation Lead and Supplemental Canister Inspection Report, CCNPP (2012).

ELECTRIC POWER RESEARCH INSTITUTE (Palo Alto, CA)

Class I Structures License Renewal Industry Report, Rev. 1, TR-103842 (1994).
Technical Basis for Extended Dry Storage of Spent Nuclear Fuel, Rep. 1003416 (2002).
Effects of Marine Environments on Stress Corrosion Cracking of Austenitic Stainless Steels, Rep. 1011820 (2005).
Climatic Corrosion Considerations for Independent Spent Fuel Storage Installations in Marine Environments, Rep. 1013524 (2006).

ERHARD, A., VÖLZKE, H., WOLFF, D., "Ageing management", paper presented at IAEA Tech. Mtg on Very Long Term Storage of Used Nuclear Fuel, Vienna, 2011.

NUCLEAR WASTE MANAGEMENT COMMISSION ESK (Bonn)

ESK Recommendations for Guides to the Performance of Periodic Safety Reviews for Interim Storage Facilities for Spent Fuel and Heat-generating Radioactive Waste (PSÜ-ZL) (2010).
Recommendation of the Nuclear Waste Management Commission (ESK): Guidelines for Dry Cask Storage of Spent Fuel and Heat-generating Waste (2013).

GILBERT, E., et al., Advances in technology for storing light water reactor spent fuel, Nucl. Technol. 89 (1990) 141–161.

HOFFMAN, E.N., SKIDMORE, T.E., DAUGHERTY, W.L., DUNN, K.A., Long Term Aging and Surveillance of 9975 Package Components, SRNL-STI-2009-00733, Savannah River National Laboratory, Aiken, SC (2009).

INTERNATIONAL ATOMIC ENERGY AGENCY (Vienna)

Storage of Water Reactor Spent Fuel in Water Pools: Survey of World Experience, Technical Reports Series No. 218 (1982).
Impact of Extended Burnup on the Nuclear Fuel Cycle, IAEA-TECDOC-699 (1993).
Spent Fuel Performance Assessment and Research, IAEA-TECDOC-1343 (2003).
Optimization Strategies for Cask Design and Container Loading in Long Term Spent Fuel Storage, IAEA-TECDOC-1523 (2006).

JOHNSON, A.B., Jr., BURKE, S.P., K Basin Corrosion Program Report, Rep. WHC-EP-0877, Westinghouse Hanford, Richland, WA (1995).

KESSLER, J., "Used fuel extended storage: What the US industry wants from DOE", paper presented at the NEI Used Fuel Management Conf. Baltimore, 2011.

KOIZUMI, S., SHIRAI, K., "Demonstration program of long-term storage (FY2004–2008): SCC of MPC under the condition of sea salt deposition", presentation to the Nuclear Regulatory Commission, 2004.

KOJIMA, S., "The new approach to regulating long-term storage of spent fuel", paper presented at the NRC Regulatory Information Conf. 2011.

KOWALEWSKY, H., et al., "Safety assessment of leak tightness criteria for radioactive materials transport packages" (Proc. 12th Int. Conf. on the Packaging and Transportation of Radioactive Materials, 1998).

KUMAR MEHTA, P., Concrete: Structure, Properties, and Materials, Prentice-Hall, Upper Saddle River, NJ (1986).

KUSTAS, F.M., et al., Investigation of the Condition of Spent Fuel Pool Components, Rep. PNL-3513, Pacific Northwest Laboratory, Richland, WA (1981).

NUCLEAR REGULATORY COMMISSION (Washington, DC)

Standard Review Plan for Spent Fuel Dry Storage Facilities, NUREG-1567 (2000).
Standard Review Plan for Transportation Packages for Spent Nuclear Fuel, NUREG-1617 (2000).
Standard Review Plan for Spent Fuel Dry Storage Systems at a General License Facility: Final Report, NUREG-1536, Revision 1 (2010).

PALMQUIST, C.A., 105-C Reactor Interim Safe Storage Project Final Report, Rep. BHI-01231, Bechtel Hanford, Richland, WA (1998).

PESCATORE, C., COWGILL, M., Temperature Limit Determination for the Inert Dry Storage of Spent Nuclear Fuel, EPRI TR-103949, Electric Power Research Institute, Palo Alto, CA (1994).

PROBST, U., et al., Investigation of seal effects according to axial compression variation of metal seals for transport and storage casks, Packag. Transp. Storage Secur. Radioact. Mater. 19 (2008) 47–52.

SASSOULAS, et al., Ageing of metallic gaskets for spent fuel casks: Century-long life forecast from 25 000-h-long experiments, Nucl. Eng. Des. **236** (2006) 2411–2417.

SELWYN, H., FINLAY, R., BULL, P., IRWIN, A., "Storage, inspection and sip testing of spent nuclear fuel from the HIFAR materials test reactor", ENS RRFM 2002 (Trans. Int. Top. Mtg on Research Reactor Fuel Management, Ghent), European Nuclear Society, Berne (2002) 114–118.

SHIRAI, K., WATARU, M., SAEGUSA, T., ITO, C., "Long-term containment performance test of metal cask" (Proc. 13th Int. High-Level Radioactive Waste Management Conf. Albuquerque, 2011).

SKIDMORE, E., Performance Evaluation of O-ring Seals in the SAFKEG 3940A Package in KAMS (U), Rep. WSRC-TR-2003-00198, Rev. 0, Savannah River Technology Center, Aiken, SC (2003).

STEELE, L.E., et al., "Neutron irradiation embrittlement of several higher strength steels", Effects of Radiation on Structural Materials (Proc. Symp. ASTM STP-426, Atlantic City, 1966), American Society for Testing and Materials, Philadelphia, PA (1967) 346–368.

VIVEKANAND KAIN, AGARWAL, K., DE, P.K., SEETHARAMAIH, P., Environmental degradation of materials during wet storage of spent nuclear fuels, J. Mater. Eng. Perfor. **9** (2000) 317–323.

VÖLZKE, H., PROBST, U., WOLFF, D., NAGELSCHMIDT, S., SCHULZ, S., "Investigations on the long-term behavior of metal seals for spent fuel storage casks" (Proc. 52nd INMM Annual Mtg, 2011).

— "Seal and closure performance in long term storage" (Proc. PSAM11 & ESREL 2012, Helsinki, 2012).

VON DER EHE, K., JAUNICH, M., WOLFF, D., BÖHNING, M., GOERING, H., "Radiation induced structural changes of (U) HMW polyethylene with regard to its application for radiation shielding" (Proc. PATRAM 2010: 16th Int. Symp. on the Packaging and Transportation of Radioactive Materials, London, 2010).

WATARU, M., SHIRAI, K., SAEGUSA, T., ITO, C., "Long-term containment test using two full-scale lid models of DPC with metal gaskets for interim storage", paper presented at 3rd East Asia Forum on Radwaste Management: 2010EAFORM, 2010.

WOLFF, D., VON DER EHE, K., JAUNICH, M., BÖHNING, M., "Performance of neutron radiation shielding material (U)HMW-PE influenced by gamma radiation" (Proc. PSAM11 & ESREL 2012, Helsinki, 2012).

Design and siting of future spent fuel storage systems

BLUE RIBBON COMMISSION ON AMERICA'S NUCLEAR FUTURE, Transportation and Storage Subcommittee Report to the Full Commission: Updated Report, BRC, Washington, DC (2012).

CODÉE, H., "Long-term storage in the Netherlands", paper presented at IAEA Tech. Mtg on Very Long Term Storage of Used Nuclear Fuel, IAEA, Vienna, 2011.

ELECTRIC POWER RESEARCH INSTITUTE, Industry Spent Fuel Storage Handbook, Rep. 1021048, EPRI, Palo Alto, CA (2010).

REACTOR SAFETY COMMISSION RSK, Recommendation of the Reactor Safety Commission (RSK): Safety Guidelines for Dry Interim Storage of Irradiated Fuel Assemblies in Storage Casks, RSK, Bonn (2001).

WASINGER, K., HUMMELSHEIM, K., GMAL, B., SPANN, H., FABER, W., "Present status and trends of spent fuel management in Germany", paper presented at IAEA Tech. Working Group on Nuclear Fuel Cycle Options Mtg, Vienna, 2007.

Spent fuel storage configurations

INTERNATIONAL ATOMIC ENERGY AGENCY, Management of Damaged Spent Nuclear Fuel, IAEA Nuclear Energy Series No. NF-T-3.6, IAEA, Vienna (2009).

KESSLER, J., WALDROP, K., "Extended used fuel storage: EPRI perspective and collaboration initiatives", Safety of Long-term Interim Storage Facilities (Proc. Workshop, Munich, 2013), OECD, Paris (2013) 63–82.

LEVIN, A., "Technical and regulatory paths forward for accelerating implementation of the Blue Ribbon Commission recommendations", paper presented at American Nuclear Society Summer Mtg, Chicago, 2012.

NICHOL, M., "Operational challenges of extended dry storage of spent nuclear fuel" (Proc. WM2012 Conf. Phoenix, 2012).

Regulatory considerations

INTERNATIONAL ATOMIC ENERGY AGENCY, Costing of Spent Nuclear Fuel Storage, IAEA Nuclear Energy Series No. NF-T-3.5, IAEA, Vienna (2009).

ROJAS DE DIEGO, J., Economics of spent-fuel storage: A description of the methodology developed by IAEA for analysing costs, Int. At. Energy Agency Bull. **3** (1990) 34–38.

Policy considerations

APARICIO, L. (Ed.), Making Nuclear Waste Governable: Deep Underground Disposal and the Challenge of Reversibility, Andra, Chatenay-Malabry (2010).

INTERNATIONAL ATOMIC ENERGY AGENCY, The Long Term Storage of Radioactive Waste: Safety and Sustainability, IAEA, Vienna (2003).

LATOURRETTE, T., LIGHT, T., KNOPMAN, D., BARTIS, J.T., Managing Spent Nuclear Fuel: Strategy Alternatives and Policy Implications, RAND Corporation, Santa Monica, CA (2010).

NATIONAL ACADEMY OF SCIENCES, Disposition of High-level Waste and Spent Nuclear Fuel: The Continuing Societal and Technical Challenges, National Academy Press, Washington, DC (2001).

OECD NUCLEAR ENERGY AGENCY (Paris)

Geological Disposal of Radioactive Waste: Review of the Development in the Last Decade (1999).
Stepwise Approach to Decision Making for Long-term Radioactive Waste Management: Experience, Issues and Guiding Principles (2004).

Other key considerations

INTERNATIONAL ATOMIC ENERGY AGENCY (Vienna)

Technical, Economic and Institutional Aspects of Regional Spent Fuel Storage Facilities, IAEA-TECDOC-1482 (2005).
Risk Management of Knowledge Loss in Nuclear Industry Organizations (2006).
Introduction to the Use of the INPRO Methodology in a Nuclear Energy System Assessment, IAEA Nuclear Energy Series No. NP-T-1.12 (2010).
Comparative Analysis of Methods and Tools for Nuclear Knowledge Preservation, IAEA Nuclear Energy Series No. NG-T-6.7 (2011).

IZURU, H., "Interface issues arising between storage and transport for storage facilities using storage/transport dual purpose dry metal casks", Management of Spent Fuel from Nuclear Power Reactors (Proc. Int. Conf. Vienna, 2010), IAEA, Vienna (2015) CD-ROM.

NATIONAL RADIOACTIVE WASTE MANAGEMENT AGENCY, 2009 Sustainable Development Report, Andra, Chatenay-Malabry (2010).

PESCATORE, C., "Preserving records, knowledge and memory over decades and more, lessons from the NEA RK&M Project", Safety of Long-term Interim Storage Facilities (Proc. Workshop, Munich, 2013), OECD, Paris (2013) 285–297.

VERHEUL, I., Networking for Digital Preservation: Current Practice in 15 National Libraries, K.G. Saur, Munich (2006).

ABBREVIATIONS

DCSS   dry cask storage system

HLW   high level waste

OECD   Organisation for Economic Co-operation and Development

SFA   spent fuel assembly

SFM   spent fuel management

SFS   spent fuel storage

SSCs   structures, systems and components

CONTRIBUTORS TO DRAFTING AND REVIEW

Bevilacqua, A.	International Atomic Energy Agency
Carlsen, B.	Idaho National Laboratory, United States of America
Chiguer, M.	Areva, France
González-Espartero, A.	International Atomic Energy Agency
Grahn, P.	Svensk Kärnbränslehantering AB, Sweden
Saegusa, T.	Central Research Institute of Electric Power Industry, Japan
Sampson, M.	Nuclear Regulatory Commission, United States of America
Seelev, I.	Rosatom, Russian Federation
Wasinger, K.	Areva, Germany
Waters, M.	Nuclear Regulatory Commission, United States of America
Wolff, D.	Bundesanstalt für Materialforschung und -prüfung, Germany

Technical Meetings

Vienna, Austria: 26–28 April 2011, 22–24 October 2012

Consultants Meetings

Vienna, Austria: 11–13 October 2010, 26–29 April 2011,
16–19 July 2012, 22–25 October 2012, 27–29 May 2013

Structure of the IAEA Nuclear Energy Series

Nuclear Energy Basic Principles
NE-BP

Nuclear General Objectives
NG-O

1. Management Systems
NG-G-1.#
NG-T-1.#

2. Human Resources
NG-G-2.#
NG-T-2.#

3. Nuclear Infrastructure and Planning
NG-G-3.#
NG-T-3.#

4. Economics
NG-G-4.#
NG-T-4.#

5. Energy System Analysis
NG-G-5.#
NG-T-5.#

6. Knowledge Management
NG-G-6.#
NG-T-6.#

Nuclear Power Objectives
NP-O

1. Technology Development
NP-G-1.#
NP-T-1.#

2. Design and Construction of Nuclear Power Plants
NP-G-2.#
NP-T-2.#

3. Operation of Nuclear Power Plants
NP-G-3.#
NP-T-3.#

4. Non-Electrical Applications
NP-G-4.#
NP-T-4.#

5. Research Reactors
NP-G-5.#
NP-T-5.#

Nuclear Fuel Cycle Objectives
NF-O

1. Resources
NF-G-1.#
NF-T-1.#

2. Fuel Engineering and Performance
NF-G-2.#
NF-T-2.#

3. Spent Fuel Management and Reprocessing
NF-G-3.#
NF-T-3.#

4. Fuel Cycles
NF-G-4.#
NF-T-4.#

5. Research Reactors — Nuclear Fuel Cycle
NF-G-5.#
NF-T-5.#

Radioactive Waste Management and Decommissioning Objectives
NW-O

1. Radioactive Waste Management
NW-G-1.#
NW-T-1.#

2. Decommissioning of Nuclear Facilities
NW-G-2.#
NW-T-2.#

3. Site Remediation
NW-G-3.#
NW-T-3.#

Key
BP: Basic Principles
O: Objectives
G: Guides
T: Technical Reports
Nos 1-6: Topic designations
#: Guide or Report number (1, 2, 3, 4, etc.)

Examples
NG-G-3.1: Nuclear General (**NG**), Guide, Nuclear Infrastructure and Planning (topic 3), **#1**
NP-T-5.4: Nuclear Power (**NP**), Report (**T**), Research Reactors (topic 5), **#4**
NF-T-3.6: Nuclear Fuel (**NF**), Report (**T**), Spent Fuel Management and Reprocessing (topic 3), **#6**
NW-G-1.1: Radioactive Waste Management and Decommissioning (**NW**), Guide, Radioactive Waste (topic 1), **#1**

40

ORDERING LOCALLY

In the following countries, IAEA priced publications may be purchased from the sources listed below or from major local booksellers.

Orders for unpriced publications should be made directly to the IAEA. The contact details are given at the end of this list.

CANADA

Renouf Publishing Co. Ltd

22-1010 Polytek Street, Ottawa, ON K1J 9J1, CANADA
Telephone: +1 613 745 2665 • Fax: +1 643 745 7660
Email: order@renoufbooks.com • Web site: www.renoufbooks.com

Bernan / Rowman & Littlefield

15200 NBN Way, Blue Ridge Summit, PA 17214, USA
Tel: +1 800 462 6420 • Fax: +1 800 338 4550
Email: orders@rowman.com Web site: www.rowman.com/bernan

CZECH REPUBLIC

Suweco CZ, s.r.o.

Sestupná 153/11, 162 00 Prague 6, CZECH REPUBLIC
Telephone: +420 242 459 205 • Fax: +420 284 821 646
Email: nakup@suweco.cz • Web site: www.suweco.cz

FRANCE

Form-Edit

5 rue Janssen, PO Box 25, 75921 Paris CEDEX, FRANCE
Telephone: +33 1 42 01 49 49 • Fax: +33 1 42 01 90 90
Email: formedit@formedit.fr • Web site: www.form-edit.com

GERMANY

Goethe Buchhandlung Teubig GmbH

Schweitzer Fachinformationen
Willstätterstrasse 15, 40549 Düsseldorf, GERMANY
Telephone: +49 (0) 211 49 874 015 • Fax: +49 (0) 211 49 874 28
Email: kundenbetreuung.goethe@schweitzer-online.de • Web site: www.goethebuch.de

INDIA

Allied Publishers

1st Floor, Dubash House, 15, J.N. Heredi Marg, Ballard Estate, Mumbai 400001, INDIA
Telephone: +91 22 4212 6930/31/69 • Fax: +91 22 2261 7928
Email: alliedpl@vsnl.com • Web site: www.alliedpublishers.com

Bookwell

3/79 Nirankari, Delhi 110009, INDIA
Telephone: +91 11 2760 1283/4536
Email: bkwell@nde.vsnl.net.in • Web site: www.bookwellindia.com

ITALY

Libreria Scientifica "AEIOU"

Via Vincenzo Maria Coronelli 6, 20146 Milan, ITALY
Telephone: +39 02 48 95 45 52 • Fax: +39 02 48 95 45 48
Email: info@libreriaaeiou.eu • Web site: www.libreriaaeiou.eu

JAPAN

Maruzen-Yushodo Co., Ltd

10-10 Yotsuyasakamachi, Shinjuku-ku, Tokyo 160-0002, JAPAN
Telephone: +81 3 4335 9312 • Fax: +81 3 4335 9364
Email: bookimport@maruzen.co.jp • Web site: www.maruzen.co.jp

RUSSIAN FEDERATION

Scientific and Engineering Centre for Nuclear and Radiation Safety

107140, Moscow, Malaya Krasnoselskaya st. 2/8, bld. 5, RUSSIAN FEDERATION
Telephone: +7 499 264 00 03 • Fax: +7 499 264 28 59
Email: secnrs@secnrs.ru • Web site: www.secnrs.ru

UNITED STATES OF AMERICA

Bernan / Rowman & Littlefield

15200 NBN Way, Blue Ridge Summit, PA 17214, USA
Tel: +1 800 462 6420 • Fax: +1 800 338 4550
Email: orders@rowman.com • Web site: www.rowman.com/bernan

Renouf Publishing Co. Ltd

812 Proctor Avenue, Ogdensburg, NY 13669-2205, USA
Telephone: +1 888 551 7470 • Fax: +1 888 551 7471
Email: orders@renoufbooks.com • Web site: www.renoufbooks.com

Orders for both priced and unpriced publications may be addressed directly to:

Marketing and Sales Unit
International Atomic Energy Agency
Vienna International Centre, PO Box 100, 1400 Vienna, Austria
Telephone: +43 1 2600 22529 or 22530 • Fax: +43 1 26007 22529
Email: sales.publications@iaea.org • Web site: www.iaea.org/books

19-00071